TOPICS IN MILLIMETER WAVE TECHNOLOGY

VOLUME 2

TOPICS IN MILLIMETER WAVE TECHNOLOGY

VOLUME 2

Edited by *KENNETH J. BUTTON*
NATIONAL MAGNET LABORATORY
MASSACHUSETTS INSTITUTE OF TECHNOLOGY
CAMBRIDGE, MASSACHUSETTS

ACADEMIC PRESS, INC.
Harcourt Brace Jovanovich, Publishers
Boston San Diego New York
Berkeley London Sydney
Tokyo Toronto

ACADEMIC PRESS, INC.
1250 Sixth Avenue, San Diego, CA 92101

United Kingdom Edition published by
ACADEMIC PRESS INC. (LONDON) LTD.
24–28 Oval Road, London NW1 7DX

Library of Congress Cataloging-in-Publication Data

Topics in millimeter wave technology.

Includes bibliographies and index.
1. Millimeter wave devices. 2. Wave guides.
I. Button, Kenneth John.
TK7876.5.T67 1987 621.381'331 87-17583
ISBN 0-12-147699-5 (vol. 1)
ISBN 0-12-147700-2 (vol. 2)

88 89 90 91 9 8 7 6 5 4 3 2 1

Printed in the United States of America

CONTENTS

CONTRIBUTORS

Numbers in parentheses indicate the pages on which authors' contributions begin.

D. L. BROWER (83), *Princeton Plasma Physics Laboratory, Princeton University, PO Box 451, Princeton, New Jersey 08544*

DENISE A. BROWN (47), *NYNEX Information Resources, Lynn, Massachusetts 01901*

N. C. LUHMANN (83), Jr., *Princeton Plasma Physics Laboratory, Princeton University, PO Box 451, Princeton, New Jersey 08544*

KAORU MOTOYA (1, 213), *Semiconductor Research Institute, Kawauchi, Sendai 980, Japan*

JUN-ICHI NISHIZAWA (1, 213), *Research Institute of Electrical Communication, Tohoku University, Sendai 980, Japan*

E. OTSUKA (173), *Department of Physics, College of General Education, Osaka University, Toyonaka, Osaka 560, Japan*

H. K. PARK (83), *Princeton Plasma Physics Laboratory, Princeton University, PO Box 451, Princeton, New Jersey 08544*

W. A. PEEBLES (83), *Princeton Plasma Physics Laboratory, Princeton University, PO Box 451, Princeton, New Jersey 08544*

MARTIN V. SNEIDER (47), *AT&T Laboratories, Crawford Hill Laboratory, Holmdel, New Jersey, 07733*

CHAPTER 1

The CW GaAs TUNNETT Diodes

Kaoru Motoya

Semiconductor Research Institute
Kawauchi, Sendai 980, Japan

Jun-ichi Nishizawa

Research Institute of Electrical Communication
Tohoku University
Sendai 980, Japan

ISBN 0-12-147700-2

I. Introduction

A semiconductor coherent signal source, such as the TUNNETT diode over 100 GHz to 1 THz (1000 GHz), has been eagerly desired for many application fields, since the IMPATT diode oscillator can not be used for the local oscillator because of its high noise and the backward wave oscillator (Carcinotron), which is a vacuum tube. The IMPATT diode oscillator has drawbacks such as a short lifetime, the need for a high voltage power supply (over 1000 V) and its large volume, including the power supply, compared to those of the semiconductor devices.

The GaAs TUNNETT diodes, which were developed by the junior author's group in the early 1950s, was introduced in Chapter 4 in Vol. 5 of this series (Nishizawa, 1982), as were the Raman semiconductor (Nishizawa, 1963, Pidgeon et al., 1971, Nishizawa and Suto, 1980) and Brillouin lasers (Suzuki et al., 1977). Far infrared generation was introduced in Chapter 6 in Vol. 7 of *Infrared and Millimeter Waves* (Nishizawa and Suto, 1983).

The avalanche multiplication phenomenon in a semiconductor has been found by the junior author (Watanabe and Nishizawa, 1952), McKay and McAfee (1953), and Gunn (1956).

The transit time negative resistance diode was proposed independently by Nishizawa and Watanabe (1953), by Shockley (1954) and by Read (1958) as a solid-state microwave source. The TUNNETT diode was proposed by Nishizawa and Watanabe (1958) as the result of analysis of the avalanching negative resistance diode which is called the IMPATT diode today.

The superiority of the TUNNETT diode to the IMPATT diode by virtue of the high frequency, low noise, and low bias voltage has been confirmed experimentally using GaAs p^+-n diodes by the J. Nishizawa's research group since 1968 (Okabe et al., 1968). At that time the existence of the TUNNETT, IMPATT and the hybrid of the TUNNETT and IMPATT modes had been determined in the course of study (Okabe et al., 1968).

The pulsed submillimeter wave oscillation of 338 GHz ($\lambda = 0.89$ mm) with 10 mW output power has been realized from GaAs p^+-n-n^+ diode (Nishizawa et al., 1979). Since then, the GaAs p^+-n^+-$i(\nu)$-n^+ has been developed in order to raise the efficiency over the p^+-n-n^+ diode by our group (Nishizawa et al., 1981, 1984).

Recently the performance limit of the IMPATT has been recognized by other workers (Elta and Haddad, 1979). The hybrid mode diode, which was named MITATT (mixed tunneling and avalanche transit time), has been developed as a CW source at 150 GHz by Elta et al. (1980); this is the first successful achievement of CW operation. This was made by the GaAs Schottky barrier diode.

It is worth mentioning here that the efficiency of tunneling and the stability for the lifetime in a *p-n* junction will be superior to that in a Shottky barrier diode, as already pointed out by Nishizawa (1976).

The poor oscillation performances from GaAs Schottky barrier type (p^+-n^+-n-n^+ diode) TUNNETT (Ohmi and Motoya, 1976) verified the above mentioned prediction with the comparison of the GaAs *p-n* junction type TUNNETT diode (Okabe et al., 1968, 1969, Nishizawa et al., 1974, Nishizawa, 1975, Nishizawa et al., 1977b, 1978a, 1978b, 1979, 1980, 1984).

Hence the research effort to realize the GaAs *p-n* junction type CW TUNNETT is thought to be valuable for many practical application fields. The object of this chapter is to present recent progress of the GaAs hyperabrupt p^+-n^+-$i(\gamma)$-n^+ TUNNETT diode. The superiority of the TUNNETT over the IMPATT and MITATT diode is also described.

II. Theory of TUNNETT Diode

A. TUNNETT Diode

The proposal of the TUNNETT diode was presented in the study of the avalanching negative resistance diode which included the diffusion effect by Nishizawa in 1958 (Nishizawa and Watanabe, 1958).

The importance of the buildup time of the avalanche injection and its spatial distribution was pointed out to determine the higher frequency limit in the IMPATT diode. This concept was developed as the avalanche induced dispersion effect by detailed numerical calculations by Nishizawa (Nishizawa, 1971, 1974, Nishizawa et al., 1974, Nishizawa et al., 1978c). The oversimplified approximation of the avalanche injection was corrected by Misawa (1966), but he did not give any physical explanation about a large effect of the time constant in the avalanche injection.

The importance of the above effect was recognized later by another researcher. The diffusion-aided spreading of the injected current pulse and the diffusion were published by several authors since 1970, after the theory was proposed by Nishizawa in 1958 (Kuvås, 1970, Gupta et al., 1975 and Schwarz, 1977). The GaAs p^+-n TUNNETT, IMPATT and MI-

TATT diodes were experimentally realized by the authors' group after 1968.

However, the millimeter wave GaAs IMPATT and MITATT diodes have poor oscillation frequency performances compared to those of GaAs TUNNETT and Si IMPATT diodes to date (Gibbons et al., 1972, Nawata et al., 1974, Schwarz and Bonek, 1978, Elta et al., 1980, Chang et al., 1981).

The influence of the tunnel injection in the transit time negative resistance diode has been presented by several authors, not including our group, after the realization of GaAs TUNNETT diode by our group since 1968 (Semichon et al., 1970, Kwok and Haddad, 1972, Chive et al., 1975, Elta and Haddad, 1978, 1979a, 1979b, Pan and Lee, 1981a and 1981b, Allen et al., 1982).

The many limitations, in the short millimeter to submillimeter wave region, of the transit time negative resistance diode, such as series resistance with decreasing device area, skin depth, the matching problem from this kind of diode to the output load and thermal resistance, are common difficulties in producing higher frequency and higher output power with a high efficiency. Their limits are common to other diodes, such as detector, mixer and varactor diodes, for short millimeter to submillimeter wave regions.

B. Small Signal Analysis of TUNNETT Diode (Okabe and Nishizawa, 1969)

The simple model of the TUNNETT diode is used, so that the voltages of the injection and transit time region are constant. The small signal analysis of the TUNNETT diode has been carried out. Fig. 1 shows the

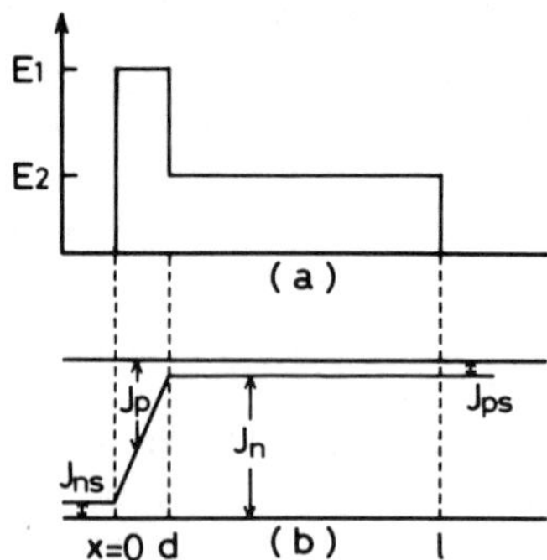

Fig. 1 (a) The electric field profile and (b) the distribution of electrons and holes in the Read-Nishizawa type TUNNETT diode.

electric field profile and the current distribution in DC condition of the Read-Nishizawa type TUNNETT diode.

C. Injection Region

Next assumptions are made that

(1) the uniform avalanche and tunnel injection under the constant electric field intensity.
(2) the saturation velocity $|v_n| = |v_p| = v_s$ for the carriers, and
(3) the neglect of the diffusion effect of carriers.

If the uniform electric field is assumed, the solving way is same as that the *p-i-n* diode analyzed by Misawa (1967). The basic equations are given by;

$$\frac{\partial E}{\partial x} = \frac{q}{\varepsilon}(N_D - N_A + p - n) \tag{1}$$

$$\frac{\partial n}{\partial t} = \frac{1}{q}\frac{\partial J_n}{\partial x} + g \tag{2}$$

$$\frac{\partial p}{\partial t} = -\frac{1}{q}\frac{\partial J_p}{\partial x} + g \tag{3}$$

and

$$J_n = -qvn \text{ and } J = -qvp. \tag{4}$$

When the uniform tunneling condition is satisfied, then

$$g = A\gamma(E) \tag{5}$$

where $\gamma(E)$ is the tunneling probability and A is the constant, respectively. The boundary conditions for $\tilde{J}_n(0)$ and $\tilde{J}_p(d)$ are given as

$$\tilde{J}_n(0) = J_{ns} \text{ and } \tilde{J}_p(d) = \tilde{J}_{ps}.$$

However, these primary currents can be neglected since the primary currents of the tunnel injection do not play as important a role as the avalanche injection.

The admittance Y is given by $\tilde{J}_n$ is divided by $\tilde{V}(d)$ and is plotted in Fig. 2. The tunneling layer width 400 Å and, $f_0 = 400$ GHz are chosen for the parameters in the calculation. The vector of the injection current does not rotate as quickly as in case of the avalanche injection. The phase delay of the injection current to the applied voltage and the attenuation of the amplitude of the injection current are small to the range of several hundreds GHz.

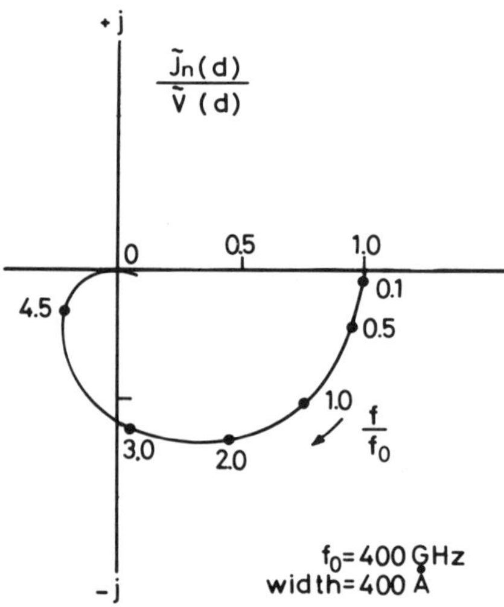

FIG. 2 Admittance of the tunnel injection region. f_0 = 400 GHz and d = 0.04 μm.

D. Drift Region ($d < x < l$)

Two assumptions, that

(a) there is no carrier generation or recombination, and
(b) the carriers drift in a saturation velocity,

are satisfied. Then the basic equations for AC components are given by

$$\frac{\partial \tilde{E}}{\partial x} = -\frac{q}{\varepsilon}\tilde{n} \tag{6}$$

$$\frac{\partial \tilde{n}}{\partial x} = -v\frac{\partial^2 \tilde{n}}{\partial x^2} + D\frac{\partial^2 \tilde{n}}{\partial x^2}. \tag{7}$$

E. No Diffusion Case ($D = 0$)

In the absence of the diffusion effect, $\tilde{n}(x)$ is given by

$$\tilde{n}(x) = \tilde{n}(d)\exp\left[-\frac{j\omega}{v}(x-d)\right]. \tag{8}$$

The electric field $\tilde{E}(x)$ is obtained by the integration of Eq. (6) using $\tilde{n}(x)$ obtained by Eq. (8). Then $\tilde{E}(x)$ is given by

$$\tilde{E}(x) = \frac{v\tilde{n}(d)}{j\omega}\left\{\exp\left[-\frac{j\omega}{v}(x-d)\right]-1\right\} + \tilde{E}(d). \tag{9}$$

The terminal voltage of $\tilde{V}(x)$ is given by integration of $\tilde{E}(x)$ of Eq. (9) as

$$\tilde{V}(x) = -\frac{v^2}{j\omega}\,\tilde{n}(d)\left\{\exp\left[-\frac{j\omega}{v}(x-d)\right] + \frac{j\omega}{v}(x-d)\right\} + \tilde{E}(d)(x-d) \tag{10}$$

The $\tilde{n}(d)$ is obtained by the injection current as using

$$\tilde{n}(d) = -\tilde{J}_n\frac{(d)}{qv}. \tag{11}$$

Then $\tilde{E}(d)$ and $\tilde{V}(d)$ are given. The total admittance Y_d across the diode is given by

$$Y_d = \frac{\tilde{J}_t}{\tilde{V}_{\text{inj}} + \tilde{V}_{\text{drift}}} = \frac{\tilde{J}_t}{\tilde{V}(l)}. \tag{12}$$

The parameter h is defined as $h \equiv l_i/l_d$ where l_i and l_d are the thicknesses of the injection and of the total layer, respectively.

The diode conductance as a parameter of h is calculated in Fig. 3. If the thickness of the injection is negligibly small with respect to the total layer

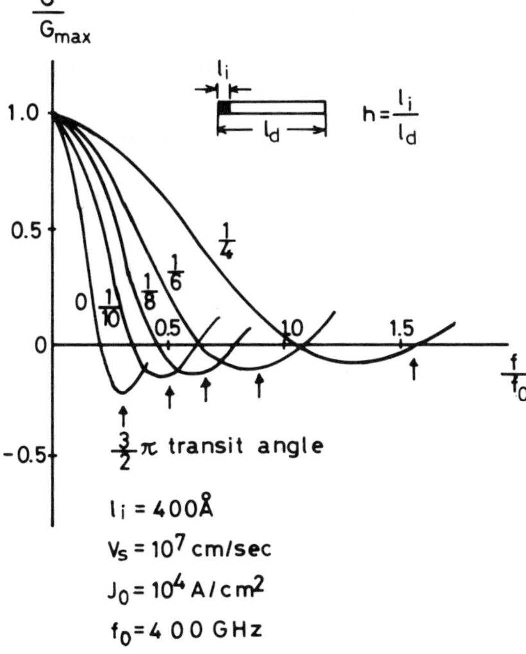

FIG. 3 Normalized conductance of the TUNNETT diode as a parameter h. $l_i = 0.04\ \mu$m, $v_s = 10^7$ cm/sec, $J_0 = 10^4$ A/cm^2 and $f_0 = 400$ GHz.

width, h is approximately zero. The maximum negative conductance with $h = 0$ is obtained at $3/2\pi$ radians transit angle (Nishizawa and Watanabe, 1958). The transit angle which gives the maximum negative conductance become smaller than $3/2\pi$ radians with h is higher than 0. And also the value of the negative conductance will be decreased with increase of h. This effect is caused by the phase delay of the injected current in the injection region. Therefore it is desirable that the width of the injection region should be minimized; the region should be as thin as possible.

F. Diffusion Effect

Boundary conditions are given as follows to solve Eq. (7).

$$\begin{aligned} \tilde{n}(x = d) &= \tilde{n}(d) \\ \tilde{n}(x = \infty) &= 0. \end{aligned} \tag{13}$$

Then $\tilde{n}(x)$ is solved to be as

$$\tilde{n}(x) = \tilde{n}(d) \exp [\lambda(x - d)]$$

where

$$\lambda = \frac{1}{2}\left[\frac{v}{D} - \sqrt{\left(\frac{v^2}{D}\right) + \frac{4j\omega}{D}}\right]. \tag{14}$$

The total AC current density is expressed as

$$\tilde{J}_t = -qv\tilde{n} + qD\frac{\partial \tilde{n}}{\partial x} + j\omega\varepsilon\tilde{E}. \tag{15}$$

The electric field intensity is given by the integral of Eq. (6) as

$$\tilde{E}(x) = -\frac{q\tilde{n}(d)}{\varepsilon\lambda}\{\exp [\lambda(x - d) - 1]\} + \tilde{E}(d) \tag{16}$$

and voltage is given by the integral of $\tilde{E}(x)$,

$$\begin{aligned} \tilde{V}(x) = &-\frac{q\tilde{n}(d)}{\varepsilon\lambda^2}\{\exp [\lambda(x - d) - 1 - \lambda(x - d)]\} \\ &+ (x - d)\tilde{E}(d) + \tilde{V}(d). \end{aligned} \tag{17}$$

The conduction current is given as

$$\begin{aligned} \tilde{J}_n(x) &= -qv\tilde{n} + qD\frac{\partial \tilde{n}}{\partial x} \\ &= -qv\tilde{n}\left\{1 - \frac{\lambda D}{v}\exp [\lambda(x - d)]\right\}. \end{aligned} \tag{18}$$

The conduction current density at the edge ($x = d$) is equal to the injection current density $\tilde{J}_n(d)$, then $\tilde{J}_n(d)$ is obtained as

$$\tilde{J}_n(d) = -qv\tilde{n}(d)\left(1 - \frac{\lambda D}{v}\right). \tag{19}$$

Then $\tilde{n}(d)$ is obtained as

$$\tilde{n}(d) = -j\frac{\tilde{J}_n(d)}{qv}\lambda. \tag{20}$$

The calculation of the admittance is shown in Fig. 4 as a parameter of diffusion constant D. With increase of D, the negative conductance decreases and finally changes to show the positive conductance.

The carrier diffusion effect plays an important part in decreasing the negative conductance (Nishizawa and Watanabe, 1958). The investigation of the diffusion constant of Si and Ge at high electric field was presented by Nishizawa et al. (Okamoto et al. 1965). They showed the increase of the diffusion constant with increase of the electric field intensity. In GaAs, it was reported that D is increased and then suddenly decreased.

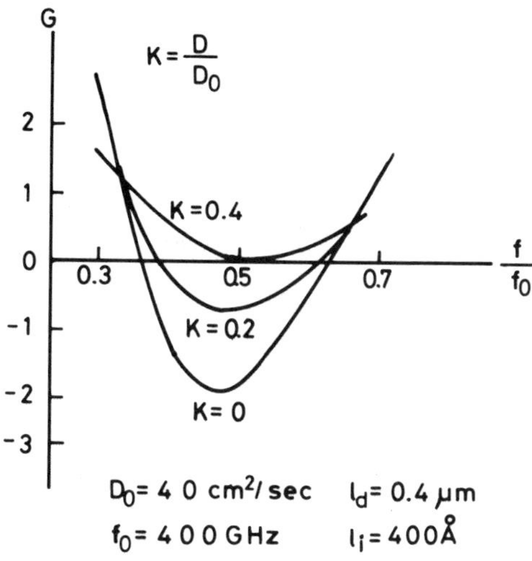

FIG. 4 Influence of the carrier diffusion effect on the conductance of the TUNNETT diode as a parameter of $K\left(\equiv \frac{D}{D_0}\right)$. $D_0 = 40\ \text{cm}^2/\text{sec}$, $f_0 = 400$ GHz, $l_d = 0.4\ \mu$m and $l_i = 0.04\ \mu$m.

G. Efficiency

If the $\pi/2$ radian injection is assumed, the efficiency of the TUNNETT is given by

$$\eta = \left| \frac{V_{\mathrm{RF}}}{V_{\mathrm{dc}}} \frac{\sin \theta_d}{\theta_d} \right| \tag{21}$$

where V_{dc} and V_{RF} are the DC bias voltage and the peak RF voltage, and θ_d is the transit angle of the drift region, respectively. More than 10% efficiency will be obtained if $|V_{\mathrm{RF}}/V_{\mathrm{dc}}|$ is 0.5 and $\theta_d = \frac{3}{2}\pi$ radians.

H. Design of the Hyperabrupt Junction p^+-n^+-i-n^+ Diode

There are several ways to obtain the p^+-n^+-i-n^+ diode, such as the epitaxial method, double diffusion and ion implantations to make the carrier profile. The hyperabrupt junction, developed to fabricate the variable capacitance diode, (Shimizu and Nishizawa, 1961) was made by the alloying diffusion method in Si material. The diffusion of the dopants during the manufacturing process is inevitable, and it is not easy to make an ideal abrupt junction.

The impurity profile of hyperabrupt p^+-n^+-$i(\nu)$-n^+ diode is considered below. The impurity profile and its electric field profile are shown in Fig. 5. The device consists of three regions as follows;

(I) hyperabrupt junction with acceptors and donors,
(II) epitaxial region with low carrier density, and
(III) substrate.

The maximum electric field intensity (E_m) should be higher than to generate tunnel injection sufficiently. In Si, E_m is over 1.2×10^6 V/cm by reported by Tyagi (1968) and in GaAs, E_m should be higher than 1.4 to 1.5×10^6 V/cm by our experimental results by p^+-n and p^+-n-n^+ GaAs TUNNETT diode (Nishizawa et al., 1968, 1978). The E_s which is the electric field intensity in the drift region is set to keep the saturation velocity (v_s) of carriers and not to occur the avalanche injection such as

$$\int_d^L \propto (E(x)) < 1.$$

The thickness of region I should be lower than the width, so as not to cause the avalanche injection. The ionization coefficient α (or β) is saturates to the value of 10^5 cm^{-1}, so the width of region I must be smaller than about 0.1 μm.

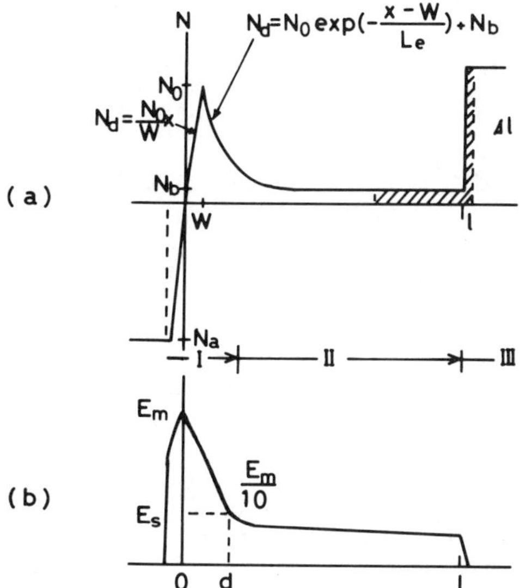

FIG. 5 (a) Impurity profile and (b) electric field profile of the hyperabrupt p^+-n^+-$i(\nu)$-n^+ TUNNETT diode.

The peak doping density N_0 of the n^+ tunnel region and the diffusion length L to obtain the tunnel injection are given as next equations. The condition for the maximum electric field intensity which is higher than the threshold electric field intensity E_{mo} is given as

$$E_{mo} \leqq \frac{q}{\varepsilon}\left\{N_0 W + LeN_0\left[1 - \exp\left(-\frac{l - W}{Le}\right)\right] + N_b(l - W) + N_s\,\Delta l\right\}. \tag{22}$$

The condition for the non avalanching in region II are given by as

$$E_s = \frac{q}{\varepsilon} N_s\,\Delta l \tag{23}$$

and

$$\alpha(E_{so})l < 1 \tag{24}$$

respectively.

If the position $x = d$ is defined as the 10% of E_m and d is less than 0.05 μm, the condition is given by next equation as,

$$\frac{9}{10} E_m \geqq \frac{q}{\varepsilon}\left\{N_0 W + LeN_0\left[1 - \exp\left(-\frac{d - W}{L}\right)\right]\right\}. \tag{25}$$

Under the reverse bias condition the region II is depleted by the next equation as

$$E_{mo} \geqq \frac{q}{\varepsilon} N_0 W + LeN_0 \left[1 - \exp\left(1 - \frac{l - W}{L_e}\right)\right]. \tag{26}$$

The range of N_0 and L_e which satisfy Eqs. (22) to (25) is plotted as shown in Fig. 6. $E_{mo} = 1.2 \times 10^6$ V/cm and $E_{so} = 2 \times 10^5$ V/cm and $W = 0.01$ μm are taken as a physical constant. The smaller L_e and higher N_0 will build the sharp tunnel injection region as can be seen in Fig. 6.

I. Calculations of the Electric Field Profile of the Hyperabrupt p^+-n^+-$i(\nu)$-n^+ Diode (Nishizawa and Motoya, 1984)

The reverse bias voltage of the p^+-n^+-$i(\nu)$-n^+ TUNNETT diode is the sum of the voltage of the tunnel injection region (V_t) and the drift region (V_d) for the injecting carriers. The ideal doping profile of the p^+-n^+-$i(\nu)$-n^+ diode is shown in Fig. 7 (a). Then n^+ region of the fabricated diode is formed by the sulfur diffusion into the $i(\nu)$ layer during the p^+ layer growth. Then the doping profile of the n^+-$i(\nu)$ layer is assumed to be the hyperabrupt junction as shown in Fig. 7 (b), accompanied by the electric field profile (Fig. 7 (c)), where the N_t is peak of the diffused n^+ layer, L is the diffusion length, N_d is the carrier density of the drift region and W_d is the thickness of the drift region, respectively.

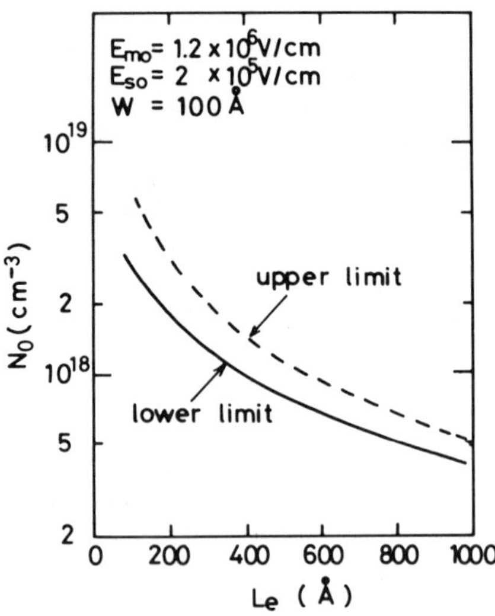

FIG. 6 Condition for the tunnel breakdown.

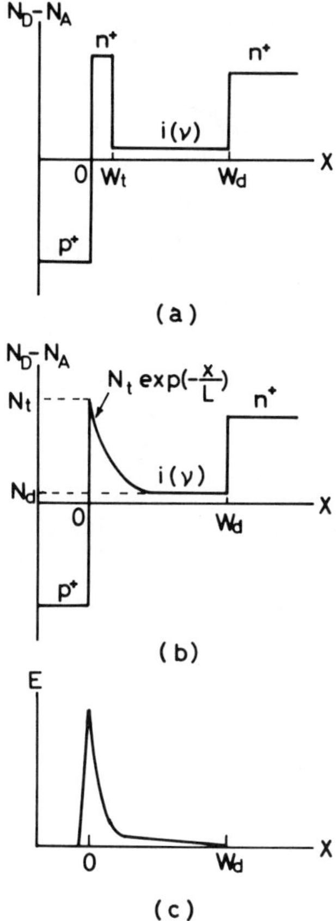

FIG. 7 (a) Ideal and (b) modified impurity profile and (c) electric field profile of the TUNNETT diode for the calculation of the electric field profile.

The impurity profile $N(x)$, the electric field intensity profile $E(x)$ and the potential at $X = W_d$ are given as follows (Sjimizu and Nishizawa, 1961).

$$N(x) = N_t \exp\left(-\frac{x}{L}\right) + N_d \tag{27}$$

$$E(x) = \left[\frac{qLN_t}{\varepsilon}\right]\left[\exp\left(-\frac{x}{L}\right) - \exp\left(-\frac{W_d}{L}\right)\right] - \left[\left(\frac{(qN_d)}{\varepsilon}\right)\right](x - W_d) - \frac{\left[\dfrac{(kT)}{(qL)}\right]\exp\left(-\dfrac{W_d}{L}\right)}{\left[\exp\left(-\dfrac{W_d}{L}\right) + \dfrac{N_d}{N_t}\right]} \tag{28}$$

$$-V(W_d) = Va + \left(\frac{kT}{q}\right)ln\left\{\left(\frac{N_tN_d}{n_i^2}\right)\left[\exp\left(-\frac{W_d}{L}\right) + \frac{N_d}{N_t}\right]\right\}$$

$$\simeq \left[\left(\frac{qN_tL^2}{\varepsilon}\right]\left[1 - \left(1 + \frac{W_d}{L}\right)\exp\left(-\frac{W_d}{L}\right) + \frac{1}{2}\left(\frac{N_d}{N_t}\right)\left(\frac{W_d}{L}\right)^2\right]$$

$$- \frac{\left(\frac{kT \cdot W_d}{qL}\right)\exp\left(-\frac{W_d}{L}\right)}{\left[\exp\left(-\frac{W_d}{L}\right) + \frac{N_d}{N_t}\right]} \tag{29}$$

where n_i is the intrinsic carrier density, q is the unit charge of an electron, ε is the dielectric constant (i.e. $\varepsilon = \varepsilon_0\varepsilon^*$), k is the Boltzmann's constant and T is the absolute temperature, respectively.

Fig. 8 shows the calculated electric field profiles using Eq. (28) as a parameter of L while N_t, N_d and W_d are fixed. When $N_t = 3 \times 10^{18}$ cm^{-3} and $N_d = 1 \times 10^{16}$ cm^{-3}, E_{max} over 1000 kV/cm and the bias voltage less than about 5V has been obtained at L is less than 300 Å as shown in Fig. 8 (a). Increasing N_t such as 1×10^{19} cm^{-3}, E_{max} increases and the bias voltage lower than 5 V has been obtained as shown in Fig. 8 (b). The small L and high N_t are needed to obtain small V_t as similar to the abrupt junction. The calculated bias voltage as a function of frequency are summarized in Table I, where $N_t = 1 \times 10^{19}$ cm^{-3}, $N_d = 1 \times 10^{16}$ cm^{-3} and L is chosen to 100 Å and 75 Å, respectively. f is given as follows,

$$f = \frac{3v_s}{4W_d} \tag{30}$$

where v_s is the saturation velocity of carriers in the drift region.

Assuming $v_s = 1 \times 10^7$ cm/sec, W_d is given as 0.25 μm at $f = 300$ GHz, 0.15 μm at $f = 500$ GHz and 750 Å at $f = 1000$ GHz, respectively. When $L = 75$ Å, bias voltage is less than 1.45 V, so there is a possibility that the bias voltage less than 1 V is necessary for the oscillation. The effect of the doping density of the drift region is investigated as a parameter of N_d in Eq. (28) as shown in Fig. 9. Increasing the N_d, E_{max} and the electric field intensity of the drift region increase, so the voltage across the diode also increases. The doping density of the drift region (N_d) affects the voltage to develop the depletion region all over the active region as shown in Fig. 9. With increase of N_d, the voltage across the diode increases and the electric field of the drift region also increases. If the ionization condition satisfies, the avalanche injection starts in a diode. The high doping density of N_t, small L and lowly doped drift region are preferable to decrease the excess voltage or electric field intensity of the drift region and to enhance

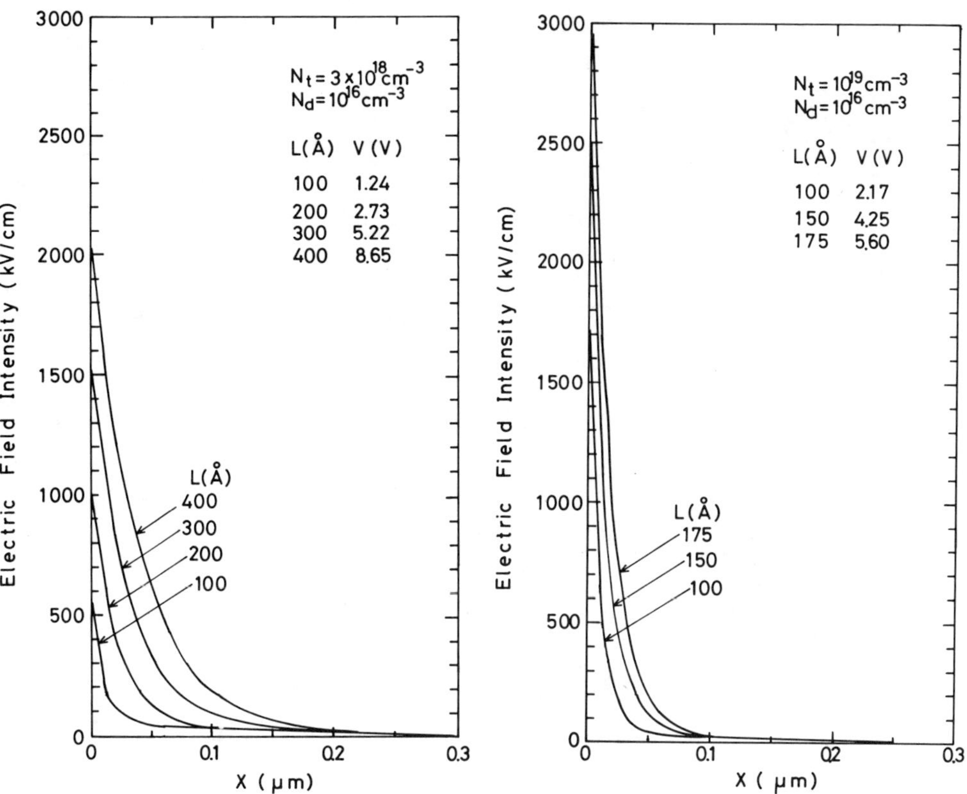

FIG. 8 The electric field profiles of the p^+-n^+-$i(\nu)$-n^+ diode. (a) $N_t = 3 \times 10^{18}$ cm^{-3}, $N_d = 10^{16}$ cm^{-3} and $W = 0.3$ μm, and (b) $N_t = 10^{19}$ cm^{-3}, $N_d = 10^{16}$ cm^{-3} and $W = 0.3$ μm.

the tunnel injection. It is needless to say that the electric field intensity of the drift region should be higher than the electric field intensity which gives the saturation velocity of carriers. If the uniform drift region with N_d and W_d is assumed, then the electric field intensity (E_d) and the needed

TABLE I

CALCULATION OF THE VOLTAGE ACROSS THE p^+-n^+-$i(\nu)$-n^+ DIODE

V / f(GHz)	$N_t = 10^{19}$cm^{-3}, $N_d = 10^{16}$cm^{-3} L = 100Å	$N_t = 10^{19}$cm^{-3}, $N_d = 10^{16}$cm^{-3} L = 75Å
300	2.17	1.45
500	1.84	1.12
1000	1.63	0.97

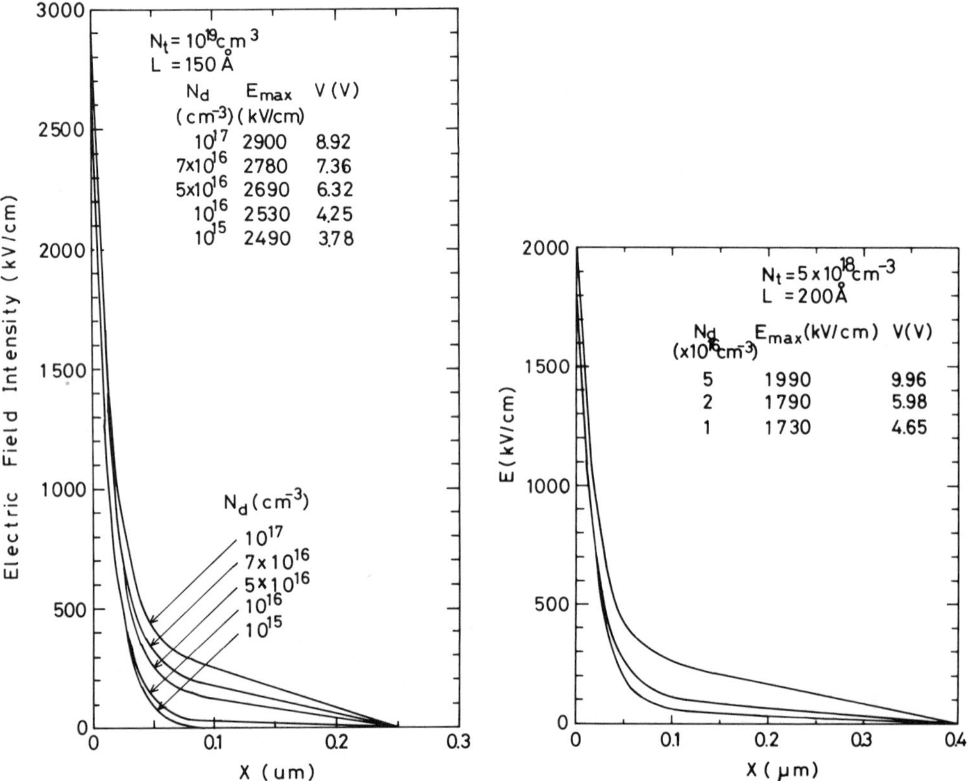

FIG. 9 The electric field profiles of the p^+-n^+-$i(\nu)$-n^+ structure as a parameter of N_d. (a) $N_t = 10^{19}\ \mathrm{cm}^{-3}$, $L = 150$ Å and $L = 0.4\ \mu$m, and (b) $N_t = 5 \times 10^{18}\ \mathrm{cm}^{-3}$, $L = 200$ Å and $L = 0.4\ \mu$m.

voltage V_d are given as follows

$$E_d = \frac{qN_D}{\varepsilon} W_d \tag{31}$$

and

$$V_d = E_d W_d. \tag{31'}$$

The relation of N_d and W_d as a parameter of E_d is shown in Fig. 10. If the space charge effect is taken into consideration, N_d is given by next equation

$$N_d = \frac{J}{qvs}. \tag{32}$$

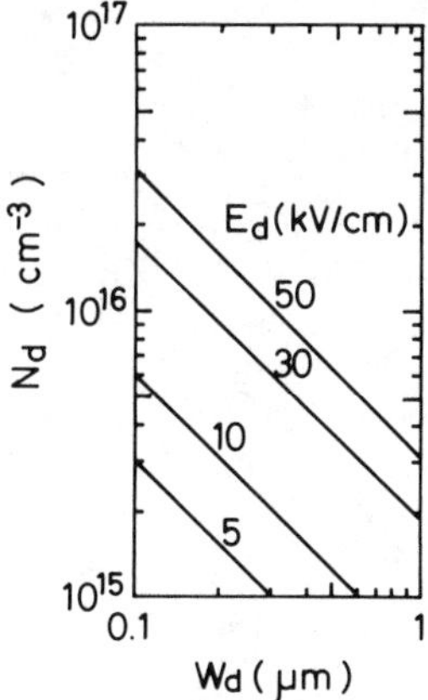

FIG. 10 The relation of N_d and W_d as a parameter of E_d to determine the doping density of the drift region.

N_d is plotted as a parameter of J and v_s as shown in Fig. 11. N_d is obtained as $6.2 \times 10^{15}\ \mathrm{cm}^{-3}$ if $J = 10^4\ \mathrm{A/cm}^2$ and $v_s = 1 \times 10^7$ cm/sec are assumed.

J. THERMAL RESISTANCE

The thermal resistance (R_θ) of the p^+-n^+-i-n^+ diode is given as follows:

$$R = \frac{1}{4aK_h} + \frac{D_g}{K_{\mathrm{Au}}\pi a^2} + \frac{W_{p^+}}{Ks\pi a^2} \tag{33}$$

where W_{p^+} is the p^+ layer thickness, D_g is the gold layer thickness from p^+-n junction to the heat sink material, K_h, K_{Au} and K_s are the thermal conductivities of a heat sink, the gold and a semiconductor and a is the junction diameter, respectively. The Eq. (33) is calculated in case of GaAs ($k_s = 0.3\,W/{}^\circ\mathrm{C} \cdot \mathrm{cm}$) by varying the junction diameter as a parameter of the heat sink material as shown in Fig. 12 (a). The diameter IIa heat sink gives the lowest thermal resistance compared to the copper heat sink. The p^+ layer thickness should be as thin as possible in order to decrease the thermal resistance. In case of the junction diameter is 20 μm, R_θ with $W_{p^+} = 1\ \mu$m is four times larger than that of R_{th} with $W_{p^+} = 0.2\ \mu$m as shown in Fig. 12 (b).

The junction temperature (T_j) of the diode is given as follows

$$T_j = I_{\mathrm{dc}} \times V_{\mathrm{dc}} \times (1 - \eta) \times R_\theta + T_a \tag{34}$$

where I_{dc} and V_{dc} are the input dc current and the bias voltage, η is the conversion efficiency from DC to RF of the diode and T_a is the ambient

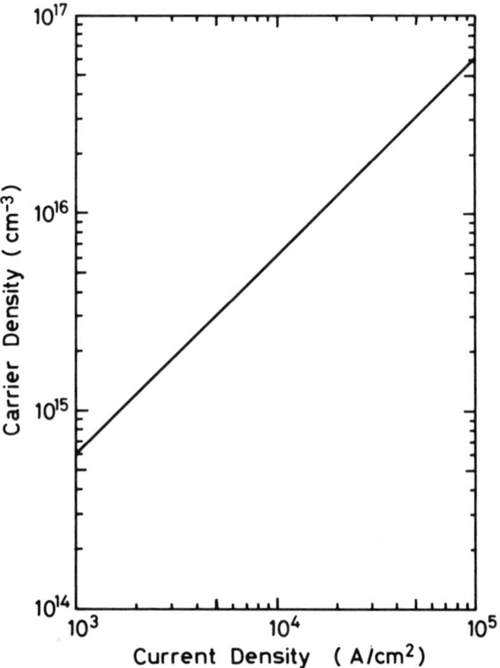

FIG. 11 The relation of N_d and J_{dc}.

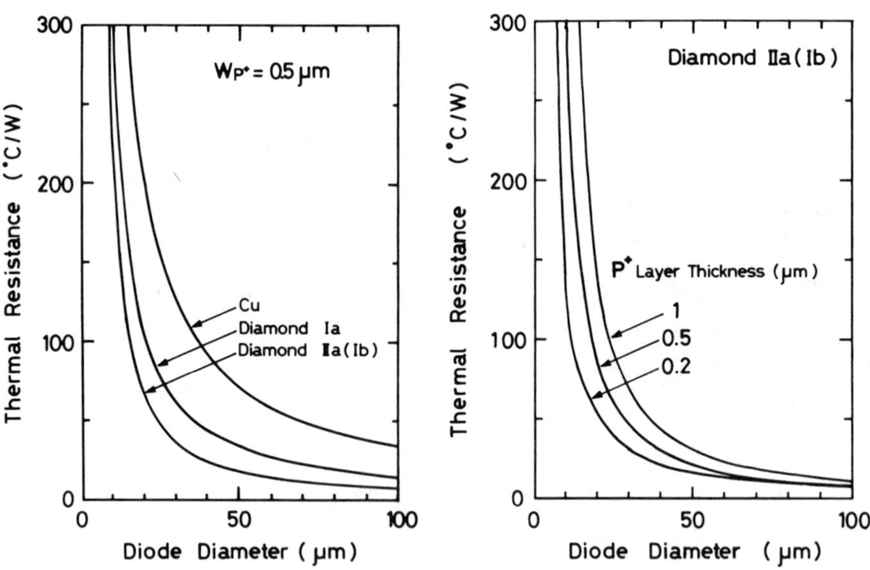

FIG. 12 Thermal resistance vs. diode diameter (a) as a parameter of heat sink materials and (b) as a parameter of p^+ layer thickness using diamond heat sink.

temperature. When the maximum operating temperature (T_m) is given, the maximum allowable input power ($\equiv I_{dc} \times V_{dc}$) can be determined from using Eqs. (33) and (34) as shown in Fig. 12 (a) and (b).

III. Preparation of the GaAs Hyperabrupt p^+-n^+-$i(\nu)$-n^+ Diode

The liquid phase epitaxy using the temperature difference method under controlled vapor pressure (TDM-CVP) has been used for the multilayer epitaxial growth onto the (100) oriented n^+ GaAs substrate. Using this method the temperature gradient is applied from melt to the substrate, so the temperature is kept constant during the epitaxial growth.

Therefore good quality of GaAs has been obtained (Nishizawa et al., 1975b, 1977a). However, the control of thin n^+ layer of the tunnel p^+-n^+ junction is difficult and the reproducibility was not good for the device fabrication.

We have adopted the growing diffusion method similar to the alloying diffusion which was successfully used for Si hyperabrupt variable capacitance diode. The Ga melt which contains Ge as for p^+ dopant and S as for n^+ dopant is used for the p^+ layer epitaxial growth. The diffusion constant of S is larger than that of Ge, so the sulfur is simultaneously diffused from growing p^+ layer to n^- layer to make thin n^+ region automatically.

The device fabrication process of the p^+-n^+-$i(\nu)$-n^+ diode was described elsewhere. The conventional structure and Au plated heat sink structure with gold-plated copper stem for pulsed operation and the diamond heat sink structure for CW operation have been developed as shown in Fig. 13 (a) and (b). The Ag/Zn/Ag multilayer structure ohmic contacts have been used to make p^+ ohmic contact in place of Ag-Zn alloy (Ishihara et al., 1976, Nishizawa et al., 1977b) and the specific contact resistance ϱ_c of this p^+ contact have been measured as low as about $8 \times 10^{-7}\ \Omega\ \text{cm}^2$ (Ema and Motoya, 1985). The diffusion length L is determined by the next equation

$$L = \sqrt{D\tau} \tag{35}$$

where D is the diffusion coefficient of the dopant and τ is the diffusion time, respectively.

The L is approximately controlled by the growth temperature (T_g) which determines the diffusion coefficient and the growth time. So in our experiment, T_g is set in the range from 750°C to 850°C. The epitaxially grown wafer is shown in Fig. 14. The obtained C–V characteristic of the p^+-n^+-$i(\nu)$-n^+ diode is shown in Fig. 15. The doping density p^+ layer is $3 \times 10^{19}\ \text{cm}^{-3}$, so the measured data shows the n^+ to n^- transition region. In this case the carrier density varies from $2.6 \times 10^{18}\ \text{cm}^{-3}$ to 5.2×10^{17}

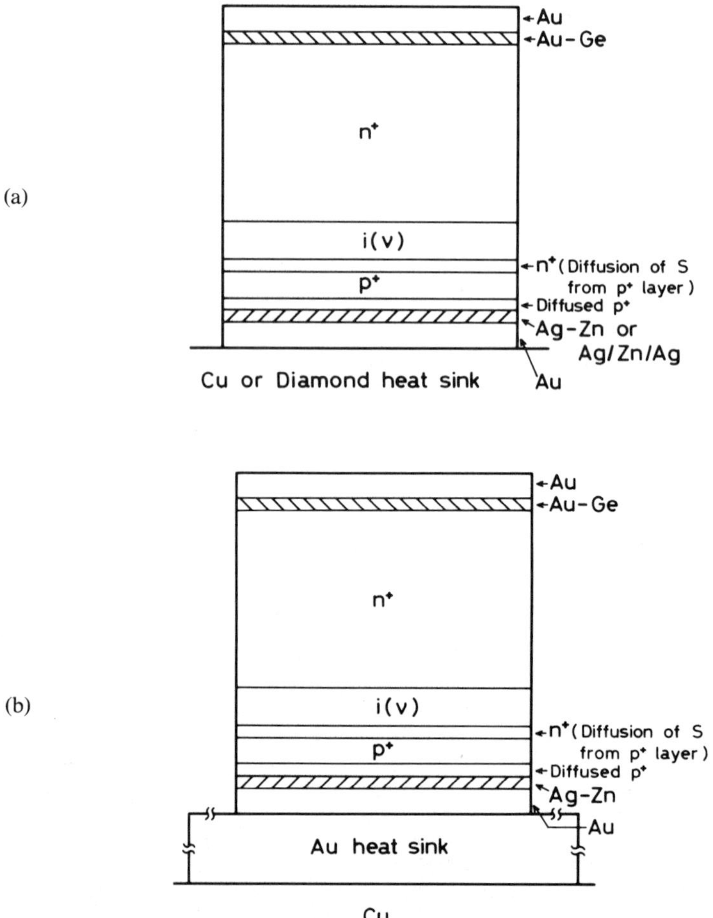

FIG. 13 Diode cross section of the p^+-n^+-$i(\nu)$-n^+ TUNNETT diode. (a) TC bonding to Cu or the diamond heat sink structure and (b) plated heatsink structure.

cm^{-3}. The diffusion length L is obtained as 100 Å from the slope of the semilog plot of the carrier profile.

Using our original TDM-CVP LPE method, p^+-n^+-$i(\nu)$-n^+ diode can be made by only two layers epitaxial growth onto the n^+ substrate. The growing apparatus is quite simple and cost effective compared to those of the molecular layer epitaxy.

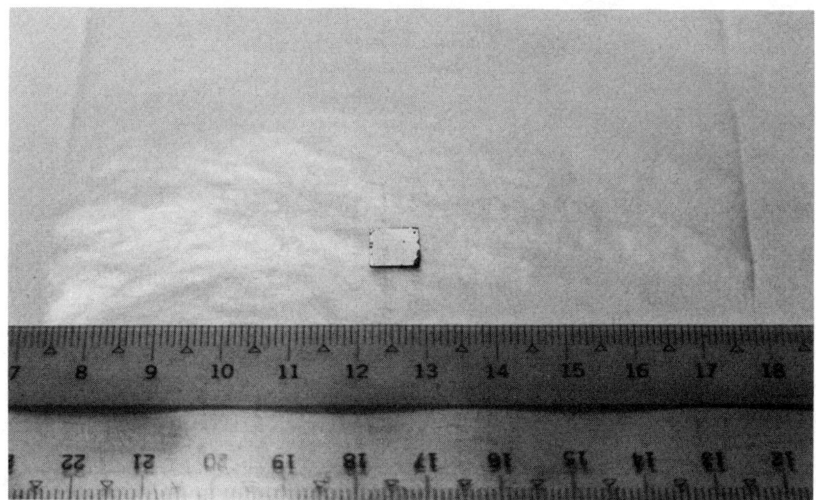

FIG. 14 The epitaxially grown wafer using the temperature difference method under controlled vapor pressure (TDM-CVP) LPE method.

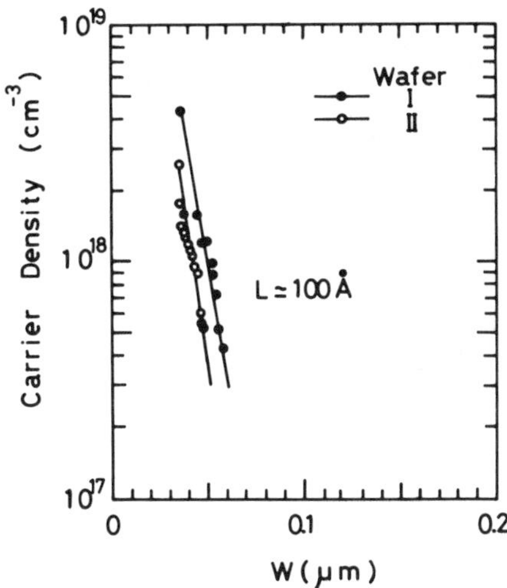

FIG. 15 Doping profile of the GaAs hyperabrupt p^+-n^+-$i(\nu)$-n^+ diodes by the C-V measurement.

IV. Experimental Results

A. *I-V* Characteristics of the Hyperabrupt GaAs p^+-n^+-$i(\nu)$-n^+ TUNNETT Diodes

The hyperabrupt p^+-n^+-$i(\nu)$-n^+ diodes in which carrier density of drift region ranges from non-doped ($10^{15} \sim 10^{16}$ cm^{-3}) to about 10^{17} cm^{-3} have been fabricated.

The relation between the current density (J) and reverse bias voltage is shown in Fig. 16. The oscillation bias points are also shown. The *I-V* curves are approximately expressed as $J \propto J_t \exp(V/V_t)$ and *I-V* characteristics do not show the sudden increase of the current of the avalanche injection from low to high current region where the oscillation occurs. The pulsed oscillation circuits are the same as published elsewhere.

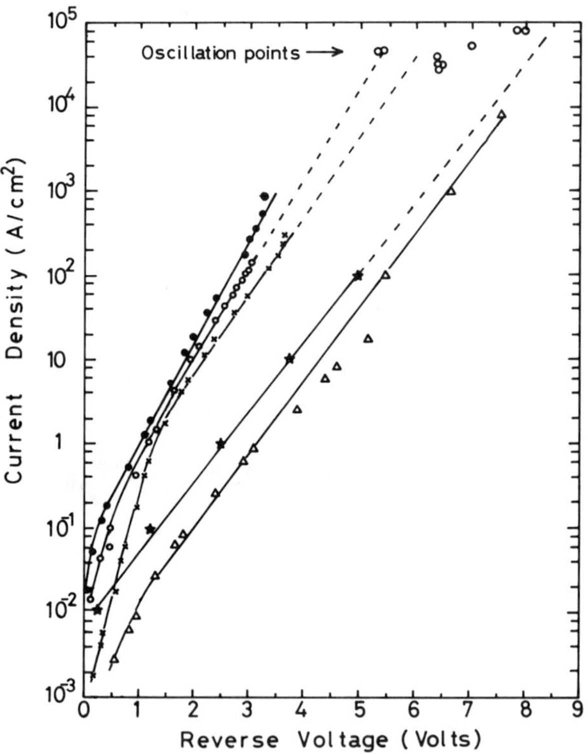

Fig. 16 *I-V* characteristics of the GaAs hyperabrupt p^+-n^+-$i(\nu)$-n^+ TUNNETT diodes by DC measurements. Oscillating bias points are also shown by the open circles.

The one of the temperature dependence of the *I-V* characteristics is shown in Fig. 17. With increase of the temperature, the current increases at the same bias voltage across the diode. This is the feature of the tunnel injection.

B. CW Oscillation from GaAs Hyperabrupt p^+-n^+-$i(\nu)$-n^+ TUNNETT Diode

The diode T-5 is mounted to the diamond heat sink as shown in Fig. 18. The diamond heat sink used here is the artificial diamond in which the thermal conductivity is the same to the natural diamond IIa ($K = 20W/\text{cm}$ °K). The top and side view of the diode can be seen. The bias circuit for the CW operation is also shown in Fig. 19. The constant current source is used to bias the diode. The resistance and the ferrite bead connected series to the diode under test are used to prevent the circuit instability in the very low frequency.

The CW oscillation characteristics are shown in Fig. 20. This is the first CW oscillation from GaAs *p-n* junction type TUNNETT diode to date. The detector voltage observed by the oscilloscope is also shown in Fig.

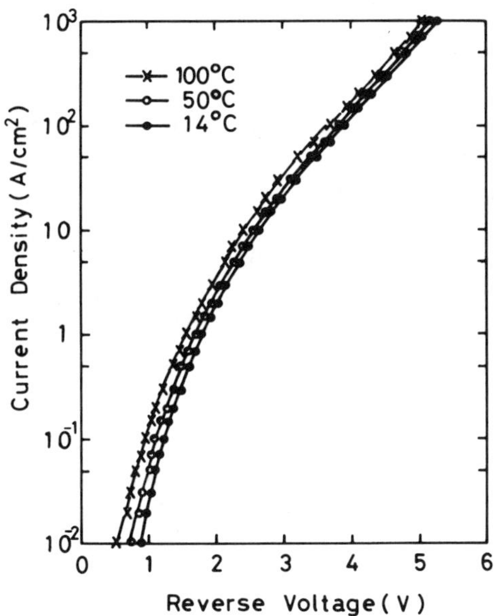

FIG. 17 Temperature dependence of the GaAs hyperabrupt p^+-n^+-$i(\nu)$-n^+ TUNNETT diode which shows the negative polarity of $\beta \left(\equiv \frac{1}{V_B}\frac{dV_B}{dT}\right)$.

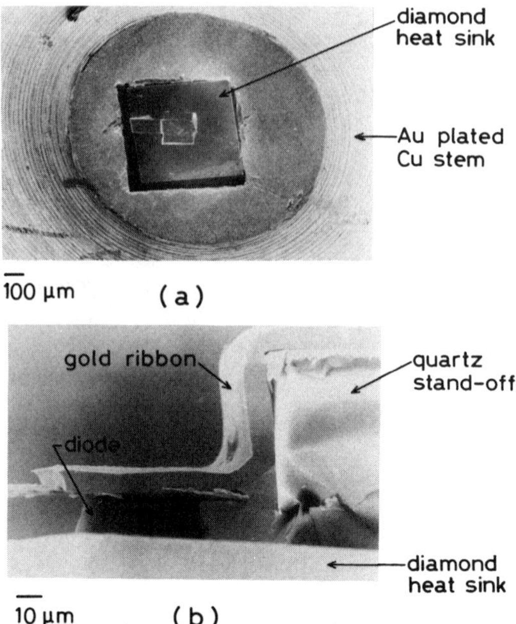

FIG. 18 SEM photograph of the CW GaAs TUNNETT diode: (a) top and (b) side view of the diamond heat sink structure.

21. The pulsed oscillation occurs from $J \sim 4 \times 10^4$ A/cm^2. However, $J_{dc} \simeq 1.5 \times 10^4$ A/cm^2 is the threshold for the CW oscillation. The decrease of the threshold current is attributed to the tunnel injection. The output power of 0.5 mW at 115 GHz with $\eta \sim 0.04\%$ from diode T-5-1. The oscillation frequency ranges from 102 to about 200 GHz. The low power output and conversion efficiency are caused by the diode series resistance. The diode thickness of 20 μm as shown in Fig. 18 is too thick to raise the diode efficiency. The 1 mW at 130 GHz output power has been obtained now.

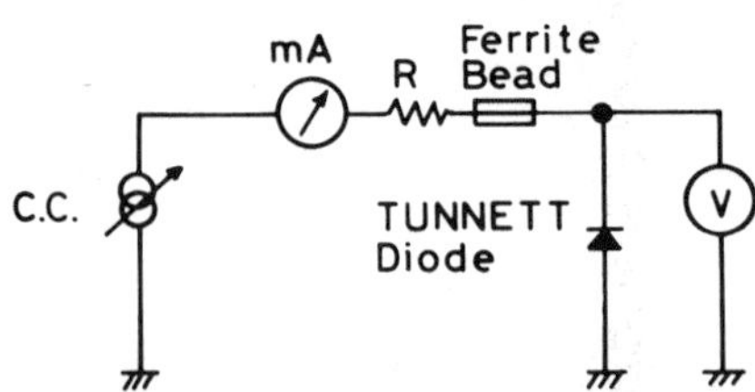

FIG. 19 Bias circuit for the CW GaAs TUNNETT diode.

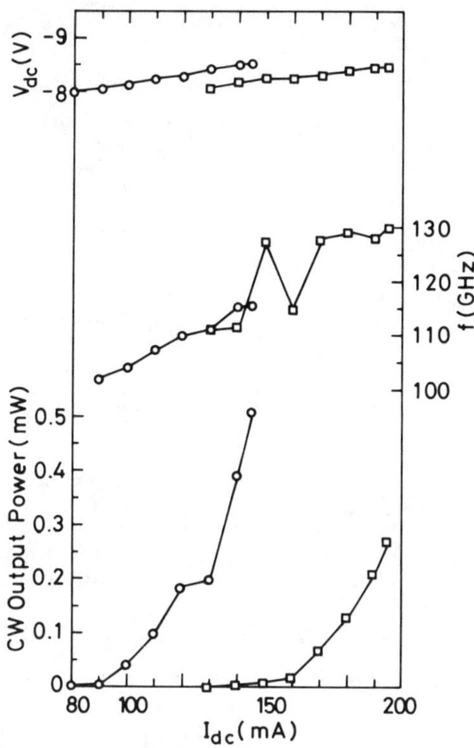

FIG. 20 Oscillation characteristics of the CW GaAs TUNNETT diode.

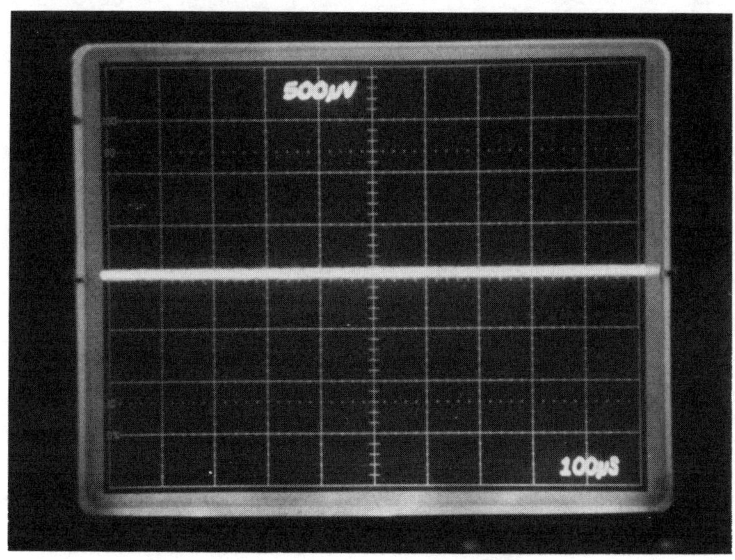

FIG. 21 Oscilloscope waveform of the CW GaAs TUNNETT diode.

The diode under CW test to date has not been optimized, so more high power, high efficiency and high frequency operation will be expected from GaAs CW p^+-n^+-$i(\nu)$-n^+ Tunnett diode.

C. Pulsed Oscillation from GaAs Hyperabrupt p^+-n^+-$i(\nu)$-n^+ TUNNETT Diode

The photograph of the pulsed oscillation circuits is shown in Fig. 22. The waveguide circuits used here are T band (110 ~ 170 GHz) and Y band (170 ~ 260 GHz).

The oscillation performance of the p^+-n^+-$i(\nu)$-n^+ TUNNETT diode is shown in Fig. 23. The oscillation occurs from $J_{th} = 4.7 \times 10^4$ A/cm² and $V_{th} = 5.2$ V and the f_{max} of 250GHz have been obtained.

In Fig. 24, the another result of the oscillation performance is given. The oscillation started from $J = 7.6 \times 10^4$ A/cm² at $f = 150$ GHz and 206 GHz oscillation was obtained at $J = 1.22 \times 10^4$ A/cm².

Up to 265 GHz oscillation has been successfully obtained as shown in Fig. 25. The 265 GHz oscillation with $J = 12 \times 10^4$ A/cm² has been obtained. The pulsed peak output power has been estimated by the output voltage of the Si point contact detector diode of T band (110 ~ 170 GHz) and the power meter (Anritzu). The calibrated diode sensitivity from 110 to 190 GHz is shown in Fig. 26. The output power 200 ~ 265 GHz has been estimated as the extrapolation of the sensitivity of the detector diode.

The peak output power and its efficiency of the pulsed p^+-n^+-$i(\nu)$-n^+ TUNNETT diodes are shown in Table I. At 130 GHz, the out power of

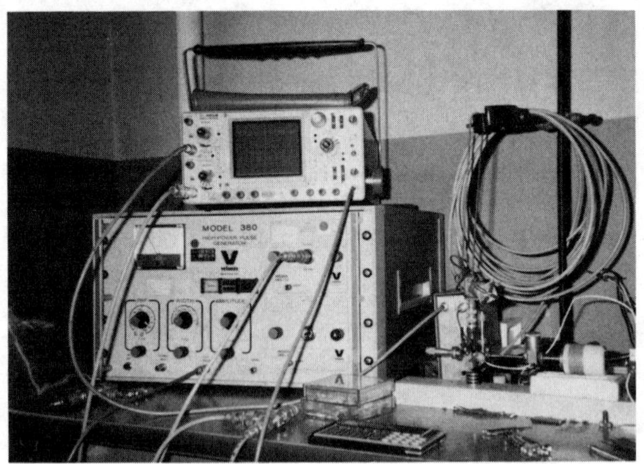

Fig. 22 Measurement system of the pulsed GaAs TUNNETT diodes.

370 mW with $\eta = 3.2\%$ efficiency has been obtained. The output power of 60 mW at 205 GHz with $\eta = 0.5\%$ and the output power of 10 mW at 260 GHz with 0.1% has been obtained, respectively.

The higher output power up to 265 GHz and up to 370 mW of 130 GHz oscillation results from the hyperabrupt p^+-n^+-$i(\nu)$-n^+ diodes are attractive for the low noise pulsed source for the radar transmitter applications. These oscillation performance over 160 GHz was obtained by using *Y* band (170 ~ 260 GHz) waveguide circuit, so the more higher power and frequency performances will be expected when these diodes are mounted in the *D* band (220 ~ 325 GHz) waveguide circuit and the introduction of new structures (Watanabe et al., 1960, Aishima et al., 1977). The influence of the temperature of these diodes has been tested by increasing the duty cycle to raise the junction temperature. The up to 1% duty cycle

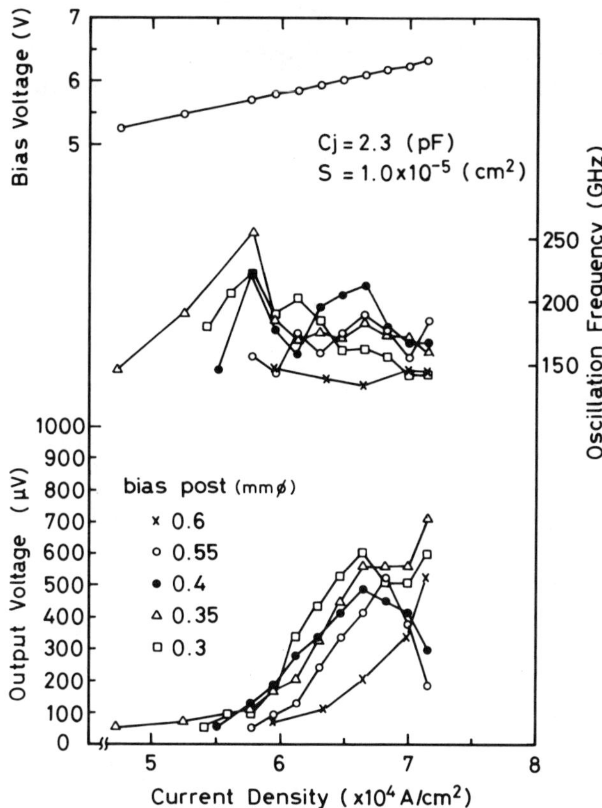

FIG. 23 Oscillation characteristics of the pulsed GaAs TUNNETT diode up to 250 GHz.

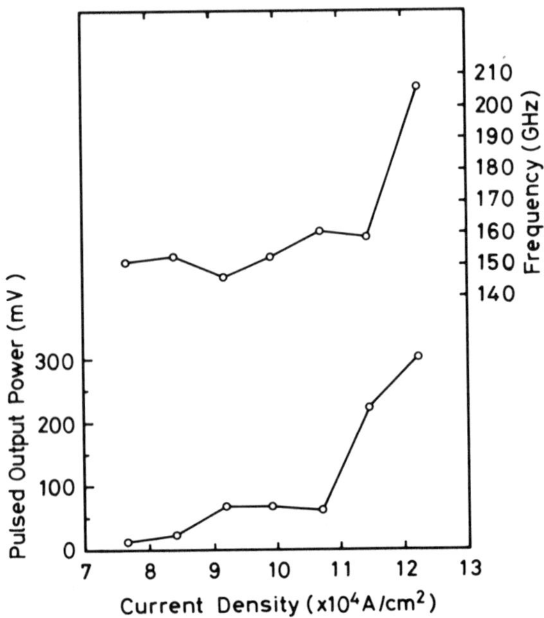

FIG. 24 Oscillation characteristics of the pulsed GaAs TUNNETT diode up to 206 GHz.

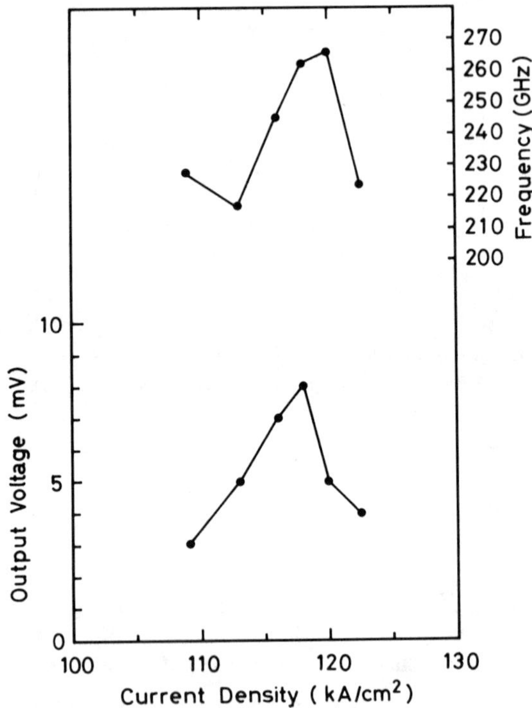

FIG. 25 Oscillation characteristics of the pulsed GaAs TUNNETT diode up to 265 GHz.

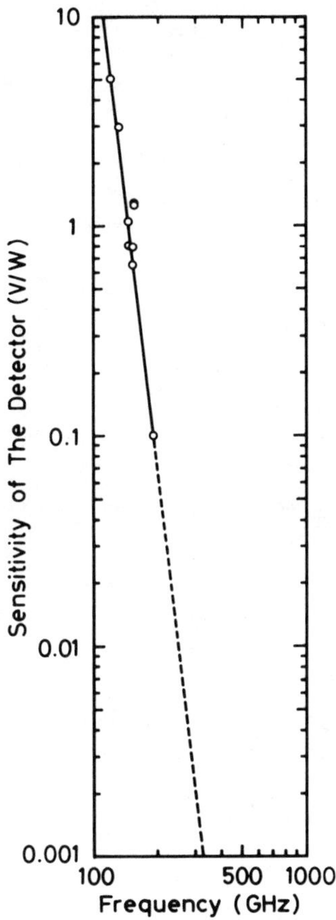

FIG. 26 Sensitivity of the detector diode.

oscillation has been obtained with high reproducibility. Characteristics are shown in Fig. 27. With increase of duty cycle, the output power increases. This is one of the important properties of the TUNNETT oscillation by virtue of its tunnel injection. From 0.1% to 1% duty cycle operation in the TUNNETT diode, the output power has been increased about 5.3dB and the frequency change is from 115.5 to 117 GHz. From 2 kHz to 10 kHz repetition rate the frequency changes within 500 MHz.

This temperature dependence to the outpower (dP/dT) is different from those of the millimeter wave Si IMPATT diode (Ishibashi et al., 1977, Kuno, 1979) and the X band GaAs pulsed IMPATT diode (Nishitani et al., 1975). Therefore the dP/dT in the oscillation performance defines the

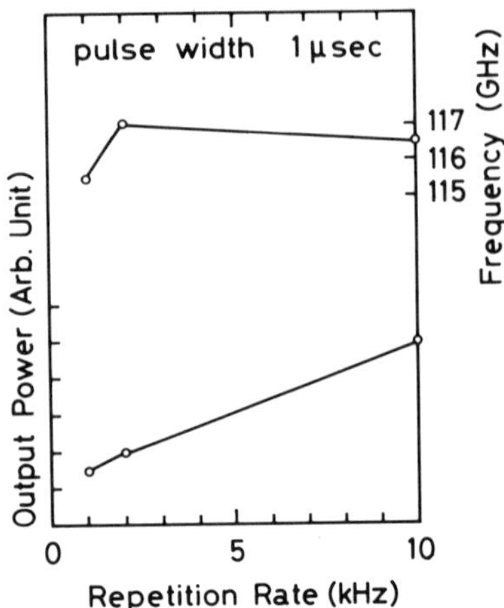

FIG. 27 Oscillation characteristics of the pulsed GaAs TUNNETT diode by varying the repetition rate.

TUNNETT and IMPATT very clearly. The summarized oscillation performances are tabulated in Table II.

D. GaAs Hyperabrupt p^+-n^+-$i(\nu)$-n^+ Diodes Operated Between TUNNETT and IMPATT Mode

Several diodes which are characterized as the hybrid of the avalanche and tunnel breakdown have been obtained during the fabrication of the pure TUNNETT diode. This kind of diode has been found by early work of our group using GaAs p^+-n diodes. An CW oscillation from GaAs Schottky barrier diode of 150 GHz 5 mW with 0.5% efficiency has been successfully obtained by Elta et al. (1980).

TABLE II

OSCILLATION PERFORMANCES FROM THE PULSED GaAs HYPERABRUPT p^+-n^+-$i(\nu)$-n^+ TUNNETT DIODES

Wafer	$C_j(V = 0V)$ ($\times 10^5$pFcm^{-2})	V_{th} (V)	J_{th} (10^4A/cm^2)	f_{max} (GHz)	Peak Power (mW)	η (%)
T − 1	2.3	5.3	4.3	254	10	0.3
T − 2	3.9	6.4	2.8	155	18	0.25
T − 3	2.15	7.0	5.65	265	10	0.1
T − 4	2.4	7.8	8.1	130	370	3.1

TABLE III

DEVICE PARAMETERS OF THE GaAs HYPERABRUPT p^+-n^+-$i(\nu)$-n^+ DIODES WITH MIXED TUNNELING AND AVALANCHE BREAKDOWN

Wafer	$C_j(V = 0V)$ ($\times 10^5$pF/cm^2)	Width (μm)		Doping (cm^{-3})		V_B(V) (at R.T.)
		p^+	n^-	$N_t(\times 10^{18})$	$N_d(\times 10^{17})$	
M − 1	1.38	0.9	0.7	0.59	1.5	9
M − 2	1.76	0.7	1.5	2.85	1.0	5.6
M − 3	2.01	0.45	0.8	2.61	1.0	6
M − 4	2.38	0.3	0.5	0.67	0.1	6.2
M − 5	2.29	0.96	0.2	0.9	0.3	5.7

The *I-V* and the millimeter wave oscillation characteristics are investigated. The device parameters are shown in Table III. The *I-V* characteristics of the diodes is shown in Fig. 28 measured at room temperature. The breakdown voltage, which can be seen by the sudden increase of the current, varies by the doping density of the n^+ and n^- layer and their thickness.

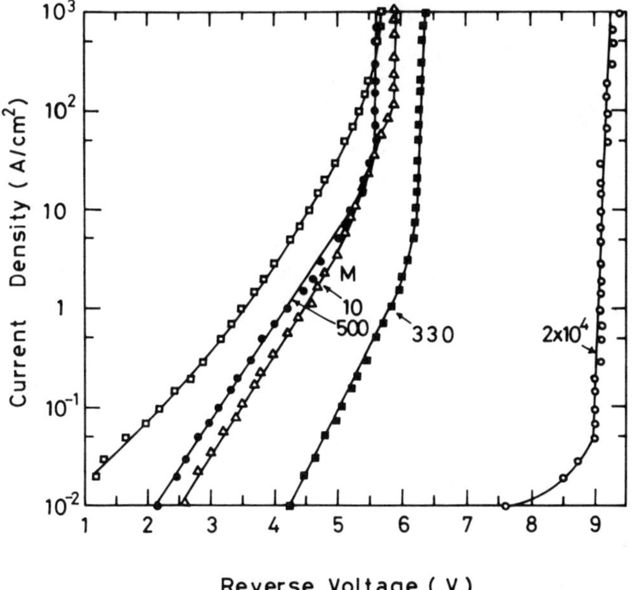

FIG. 28 DC *I-V* characteristics of the GaAs p^+-n^+-$i(\nu)$-n^+ diodes with the hybrid or mixed tunnelling and avalanche breakdown. The avalanche multiplication factor (*M*) change from large to small value with decrease of the breakdown voltage is clearly shown.

The exponential characteristics inherent to the tunnel injection have been observed in the low current region. This tendency is enhanced by the low breakdown voltage diode, since the maximum electric field at the p^+-n^+ junction is higher than that of the diode which has high breakdown voltage. The avalanche multiplication factor is defined as follows;

$$M = \frac{J}{J_{\text{sat}}} = \frac{I}{I_{\text{sat}}}.$$

The multiplication factor M, under the condition that $J = 10^3$ A/cm^2 and J_{sat} is taken as the current density just before the avalanche breakdown occurs is also shown in Fig. 28.

The temperature dependence on the millimeter wave performances has been measured by heating the waveguide cavity where the diode is mounted, as shown in Fig. 29. The diode is pulsively biased with a short pulse width of 100 nsec and 100 Hz repetition rate in order not to raise the junction temperature.

The temperature dependence of the millimeter wave oscillation performances is shown in Fig. 30 with its I-V characteristics. The temperature dependence of I-V characteristics of this diode is nearly equal to zero when measured from the room temperature to 100°C. The backshort and E-H tuner of the oscillator circuit are adjusted to obtain the maximum output power by varying the bias current.

The output power of this diode in which the ambient temperature is 21°C and 50°C is nearly the same as shown in Fig. 31. However, with increase of the ambient temperature to 100°C the threshold current for the oscillation is lowered from 1.05 A at 21°C to 0.9 A at 100°C and the output power has been increased very much.

The output power at 1.5 A bias level is three times (4.8 dB) larger than that of 21°C operation. The dependence of the output power to tempera-

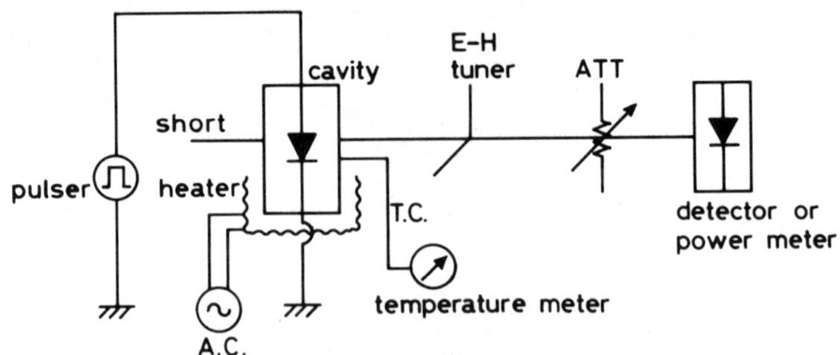

FIG. 29 Measurement circuit for varying the temperature of the pulsed oscillation circuit.

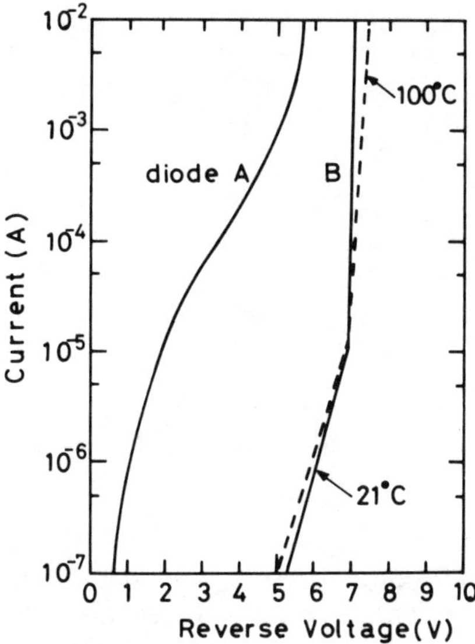

FIG. 30 Temperature dependence of the *I-V* characteristics of the hybrid or mixed tunneling and avalanche breakdown diode.

ture ($dP/dT > 0$) is the same for the characteristics of the TUNNETT diode, so the tunnel injection plays import role in the millimeter wave region to raise the device efficiency. In Fig. 32, dP/dT is also measured with use of diode B in which *I-V* characteristics is shown in Fig. 30. In this

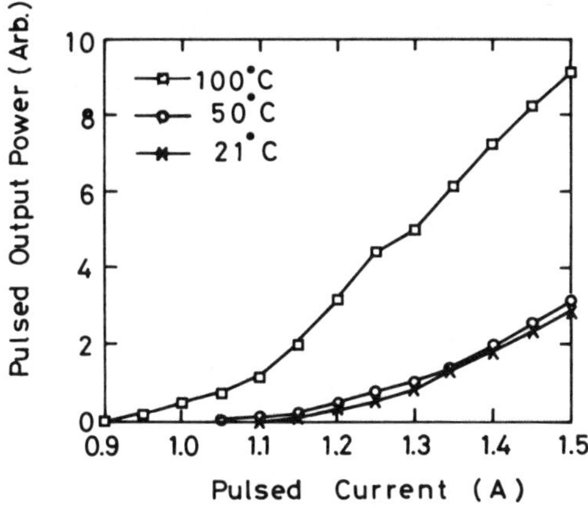

FIG. 31 Temperature dependence of the pulsed oscillation characteristics of the diode *A* as shown in Fig. 30.

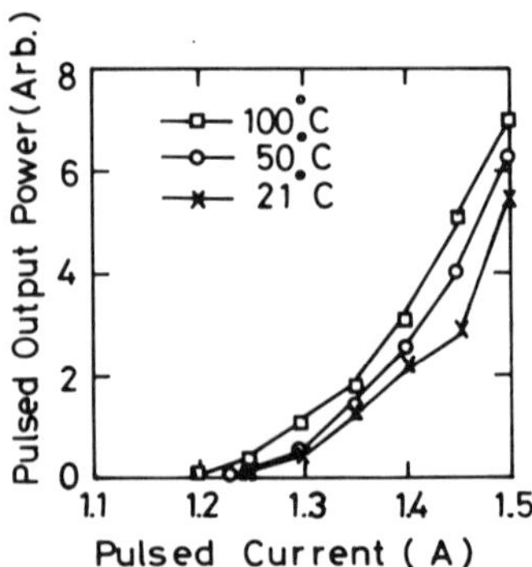

FIG. 32 Temperature dependence of the pulsed oscillation characteristics of the diode *B* as shown in Fig. 30.

case, a little output power increase and lower threshold current for the oscillation can be seen. The output power at 1.5 A bias level is 1.27 times (1.05 dB) larger than that of 21°C operation. Even in the diode where the avalanche current flows dominantly at high current level the tunnel injection also improves the device performances.

The check of the temperature dependence for both DC *I-V* characteristics and millimeter wave performance is necessary to evaluate these type diodes. The maximum oscillation frequency of the diodes is about 130 GHz as shown in Table IV. This corresponds to the maximum oscillation frequency from the abrupt p^+-n-n^+ diode in which n layer density is 5×10^{17} cm^{-3} and the temperature dependence of the voltage (β) changed from negative to zero after the avalanche injection. Much higher oscillation frequency up to 338 GHz and the higher output power have been obtained from GaAs p^+-n-n^+ or p^+-n diodes from our previous results.

TABLE IV

OSCILLATION PERFORMANCE FROM THE DIODES LISTED IN TABLE III

Wafer	$C_j(V = 0V)$ ($\times 10^5$pF/cm^2)	V_{th} (V)	J_{th} (10^4A/cm^2)	f_{max} (GHz)	Peak Power (mW)	η (%)
M − 1	1.38	9.6	2.2	115	0.05	0.001
M − 2	1.76	6.8	10	128	1.9	0.024
M − 3	2.01	6.8	9	119	8	0.082
M − 4	2.38	11.9	7.6	128	1.8	0.012
M − 5	2.29	10.8	6.8	121	1.8	0.015

The GaAs Schottky barrier IMPATT with Read-Nishizawa type structure of 5 mW 130 GHz CW with 0.5% efficiency was reported by K. Chang et al. of Hughes Aircraft Company (1981). They used the T band (110 ~ 170 GHz) circuit.

These results obtained by us and by the Hughes group may suggest that the maximum oscillation frequency of the GaAs IMPATT exists near 130 GHz.

E. MODE CHANGE FROM THE TUNNETT TO IMPATT MODE

More than 300 diodes of over 10 different wafers of the p^+-n^+-$i(\nu)$-n^+ diodes of hybrid or mixed tunnelling and avalanche injections have been tested.

However, the diodes in which the temperature dependence of the output power is nearly equal to zero ($dP/dT \simeq 0$) are scarce to date. The mixed of two oscillation frequencies from these diodes have been obtained sometimes by tuning the circuit. One of the such cases is shown in Fig. 33 as the I-V characteristics with the temperature dependence and in Fig. 34 as the oscillation characteristics. The temperature coefficient of the breakdown voltage (β) changes from negative to positive. This diode

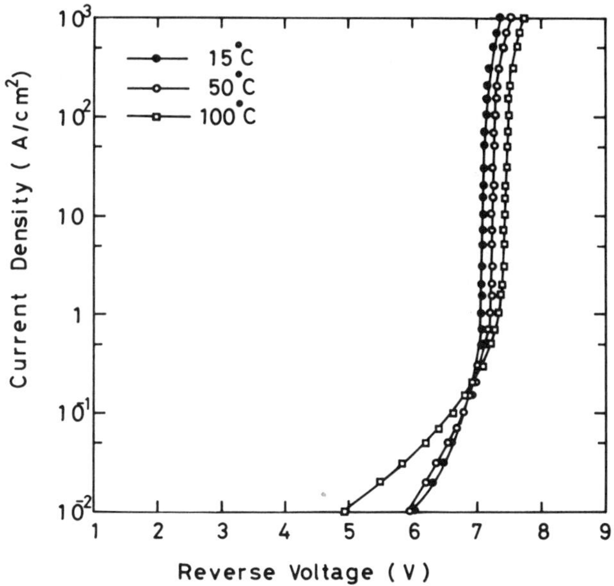

FIG. 33 Temperature dependence of the I-V characteristics of the mixed tunneling and avalanche breakdown diode.

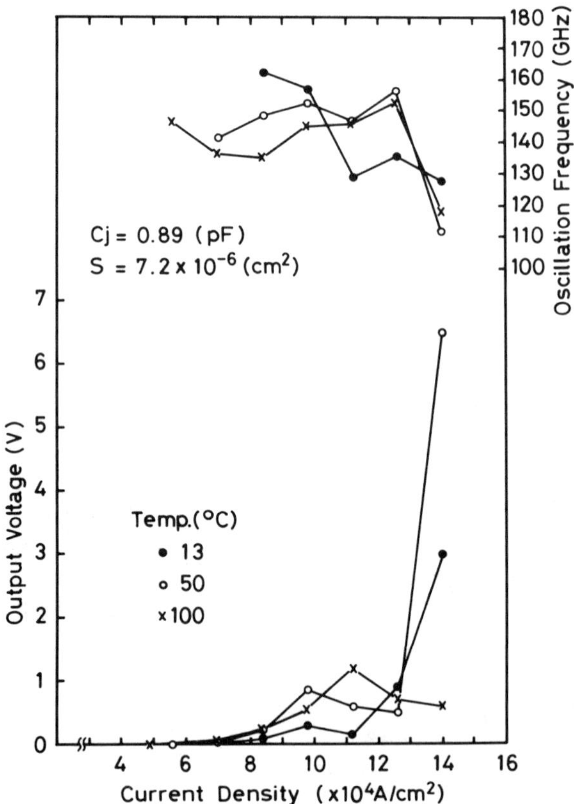

FIG. 34 Temperature dependence of the pulsed oscillation characteristics from the diode as shown in Fig. 33.

was oscillated from 130 to 162 GHz under the bias current density from 4.9 to 12.5×10^4 A/cm^2, and over this current level the sudden decrease of the oscillation frequency from 111 to 128 GHz was obtained, as shown in Fig. 34.

The ambient temperature was varied at the same time in the pulsed oscillation. The higher oscillation frequencies up to 130 to 162 GHz have been obtained at the cavity temperature of 100°C and the lower oscillation frequencies from 111 to 128 GHz also has been obtained. In this low frequency region 128, 118 and 111 GHz at 13°C, 100°C and 50°C have been obtained respectively, and the output power at 100°C was the lowest compared to those of 50°C and 13°C operations. These sudden oscillation frequency changes can be interpreted to mean that the oscillation mode changes from the TUNNETT to IMPATT mode. This phenomenon was observed more significantly by lower breakdown voltage diode as shown in Fig. 35 and Fig. 36.

The pulsed oscillation (100 nsec and 100 Hz) at 17°C occurs at the current density (J_{th}) from 12×10^4 A/cm^2 at 100 GHz. When the cavity temperature is increased from 50 to 100°C the current density threshold is decreased suddenly to 6×10^4 A/cm^2, the oscillation frequency ranges from 142 to 156 GHz and dP/dT is positive. The frequency change caused by varying the temperature was apparently small as shown in Fig. 36 at the same bias current level, so this significant oscillation change was caused not only by the temperature effect of the carrier velocity but also by the tunnel injection to the avalanche injection itself.

V. Features of the GaAs Hyperabrupt p^+-n^+-i-n^+ TUNNETT Diodes

A. FREQUENCY

The summarized oscillation performances of the hyperabrupt p^+-n^+-$i(\nu)$-n^+ TUNNETT diodes and hybrid or mixed mode of TUNNETT and IMPATT are shown in Fig. 37.

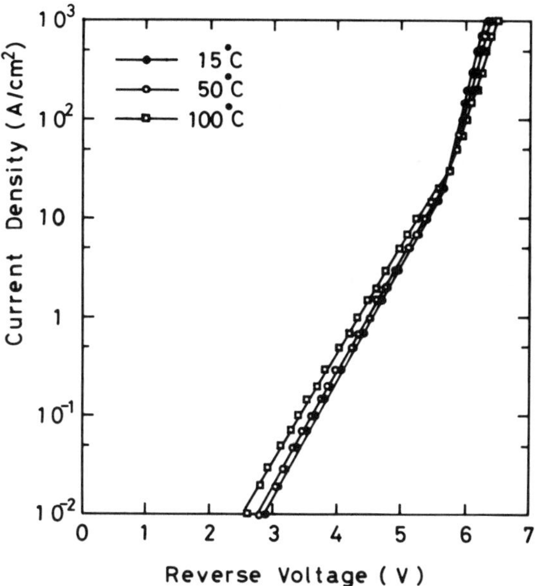

FIG. 35 Temperature dependence of the I-V characteristics of the mixed tunneling and avalanche breakdown diode.

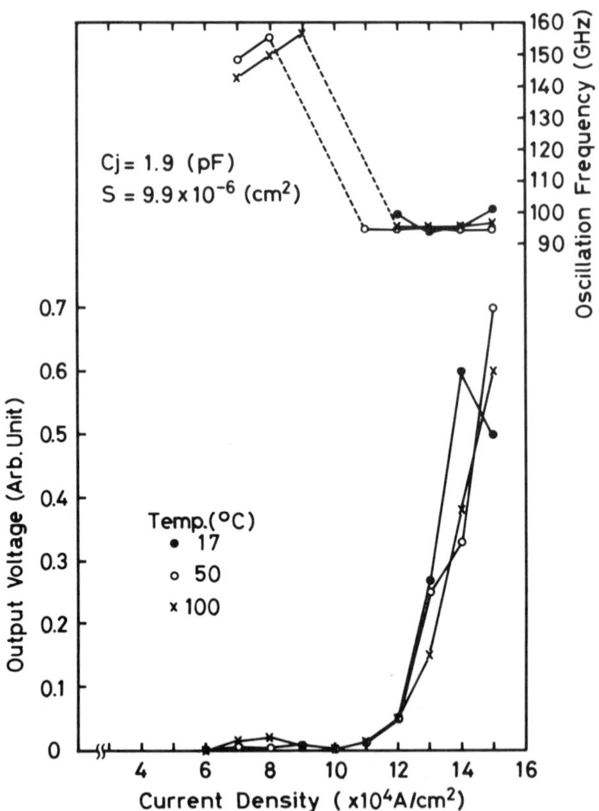

FIG. 36 Temperature dependence of the pulsed oscillation characteristics from the diode as shown in Fig. 35.

The oscillation frequency ranges from 110 to 265 GHz by the TUNNETT mode and as high as 130 GHz by the hybrid or mixed mode. Up to 338 GHz oscillation and 301 GHz oscillation from p^+-n-n^+ and the former p^+-n^+-$i(\nu)$-n^+ diodes have been obtained (Nishizawa et al., 1979 and 1980).

These experimental results clearly show the superiority of the TUNNETT diode over the IMPATT and MITATT diodes.

B. Output Power and Efficiency of the GaAs Hyperabrupt p^+-n^+-$i(\nu)$-n^+ TUNNETT Diode

The relation of the output power and the efficiency versus frequency generated by hyperabrupt p^+-n^+-$i(\nu)$-n^+ diodes is shown in Figs. 38 and 39

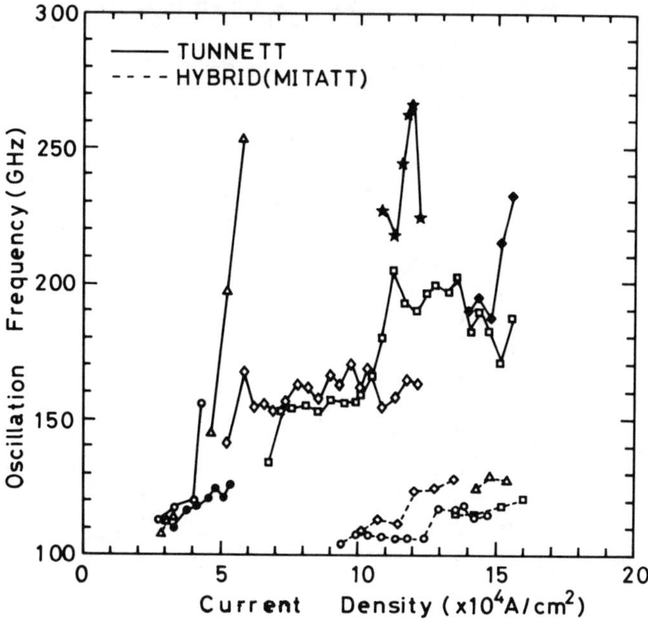

FIG. 37 Pulsed oscillation frequency versus current density from the TUNNETT and mixed mode diode.

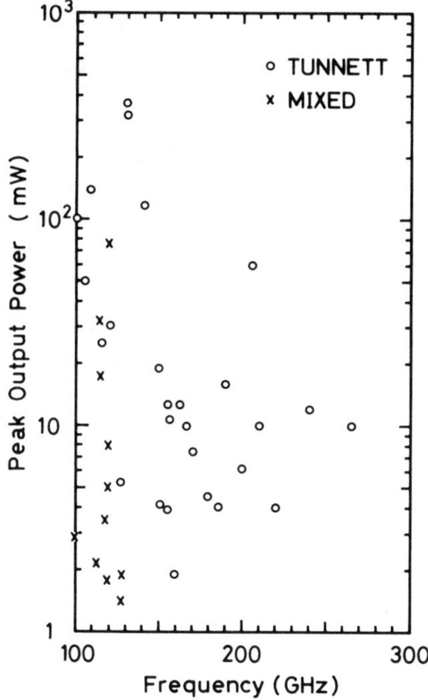

FIG. 38 Peak output power from the pulsed oscillation versus frequency from the TUNNETT and mixed mode diodes.

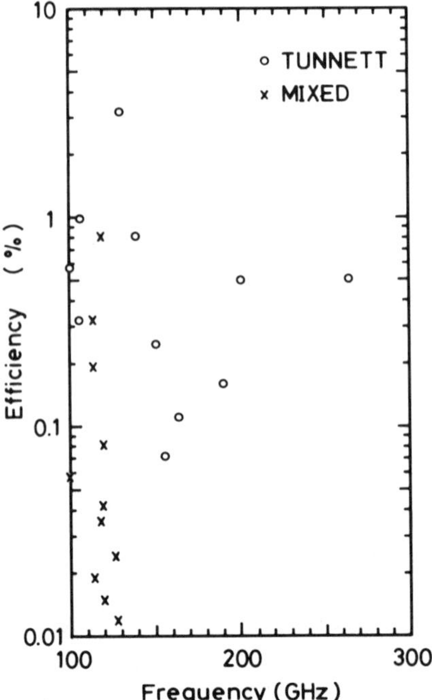

FIG. 39 Efficiency from the pulsed oscillation versus frequency from the TUNNETT and mixed mode diodes.

respectively. The higher oscillation frequency, the output power and the efficiency of the TUNNETT mode diodes are all superior to those of the IMPATT and MITATT modes diodes. These experimental results are the same conclusion of the p^+-n and p^+-n-n^+ TUNNETT diodes as published by our group since 1968.

The diodes under tests are made by the original TDM CVP LPE method and do not use the expensive apparatus such as the ion implanter, the molecular beam epitaxy (MBE), the high accuracy vapor phase (VPE) or the metal organic (MO) CVD. Therefore, the fabrication of the hyper-abrupt TUNNETT by the TDM-CVP LPE method seems to be very useful and cost effective for the practical application fields.

C. MODES OF TUNNETT, HYBRID OF TUNNETT AND IMPATT (MITATT) AND IMPATT DIODES

The criteria to determine how the operation mode from TUNNETT to IMPATT mode significantly occurs, obtained by our group, are summarized and tabulated in Table V.

TABLE V

SUMMARY OF THE OSCILLATION CHARACTERISTICS FROM THE EXPERIMENTAL RESULTS OF THE TUNNETT DIODE

TEST	DC	RF			
MODE	$\beta = \frac{1}{V_B}\frac{dV_B}{dT}$	$\frac{dP}{dT}$	$\frac{1}{I_{th}}\frac{dI_{th}}{dT}$	$\frac{f_{max}}{130GHz}$	f_{max} (GHz)
IMPATT	+	−	+		
HYBRID	− to 0	+	−	<1	130
(or MITATT)	0	0	0		
TUNNETT	− to 0 −	+	−	>1	≥1000

The temperature dependence β of the *I-V* should be measured from low to high current density level up 1×10^5 A/cm^2 is recommended primary how the β is negative or positive. Next step is the pulse driven oscillation tests which gives the f_{max}, dP/dT and $\frac{1}{I_{th}}\frac{dI_{th}}{dT}$ by using short pulse width and low repetition rate rate such as 100 Hz by varying the ambient temperature of the cavity of the diode under test.

Roughly speaking, it is not so easy to fabricate the hybrid mode to obtain the higher oscillation frequency by our extensive experiments and easy to obtain the TUNNETT mode oscillation by using p^+-n, p^+-n-n^+ and hyperabrupt p^+-n^+-$i(\nu)$-n^+ diodes. The increase of the avalanche injection occurred in the TUNNETT diode will generate more noise than that of the pure TUNNETT mode.

Therefore the hybrid or mixed mode will be inferior to the TUNNETT mode, as predicted by Nishizawa since 1958. Therefore, it is necessary to suppress the avalanche injection in the TUNNETT diode. The onset of the avalanche injection in the TUNNETT diode is not desirable, since it generates much noise to degrade the oscillation spectrum of the TUNNETT diode and is harmful for local oscillator applications.

VI. Future of the TUNNETT Diode

The tunneling injection is a quantum mechanical phenomenon and its speed is very quick. This was pointed out by Nishizawa that the carriers by tunnel injection drift with more higher speed of the saturation velocity v_s.

If non-scattering occurs in the drift space, the carrier velocity will be higher than the saturation velocity. This is the ideal mode of the TUN-

NETT diode similar to the ideal SIT (or ballistic SIT) (Nishizawa et al., 1975a, Nishizawa and Yamamoto, 1978, Nishizawa, 1979, 1980, Shur and Eastman, 1979, Bozler and Alley, 1980). The optimum frequency for the $\frac{3}{2}\pi$ radians transit angle is given by a well-known equation as

$$f = \frac{3v}{4W} \tag{36}$$

where v is the non-saturation velocity without scattering. This is shown in Fig. 40 as a parameter of v. 1 THz and 3.8 THz oscillations with $W = 750$ Å and 200 Å with $v = 10^7$ cm/sec will be expected and more higher oscillations up to 10 THz will be obtained under the condition that v is higher than 3×10^7 cm/sec and W is smaller than 300 Å.

The application of the photo-excited molecular layer epitaxy (PMLE) developed by our group (Nishizawa et al., 1984) to TUNNETT diode fabrication is a suitable epitaxy method to realize the submillimeter wave oscillation from TUNNETT up to the range of the semiconductor Raman and Brillouin lasers, since the layer by layer growth with an atomic accuracy at very low temperature such as 300°C has been already realized.

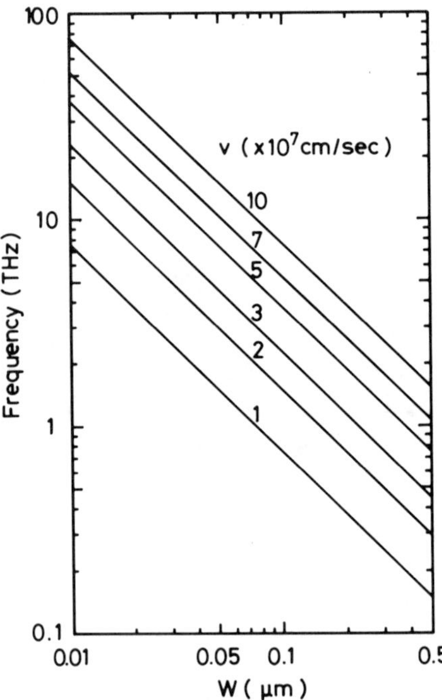

FIG. 40 Oscillation frequency versus drift region width as a parameter of the velocity under non-scattering condition.

VII. Conclusion

The CW and pulsed oscillations from GaAs hyperabrupt p^+-n^+-$i(\nu)$-n^+ diodes are presented and the experimental oscillation performances of the TUNNETT show superiority to those of the IMPATT diodes.

There is a possibility of obtaining more higher frequency oscillation over 1000 GHz by using more sophisticated device fabrication technology in the future.

References

Aishima, A., Suzuki, S., Yokoo, K., and Ono, S., (1977). 5 ~ 20 GHz wide band GUNN diode oscillator, *Conv. Record Semicond. Sect. J.E.C.E.* pp. 6–159 (in Japanese).

Allen, F., Bae, Y., Iyer, S. S., Jou, C., Kupiszewski, A., Lee, N., Luhmann, N. C. Jr., McDermott, D. B., Metzer, R. A., Pan, D. S., Peebles, W. A., Rutledge, D., Umstadter, D., Xu, Y., and Yang, C. C., (1982). Current millimeter and sub-millimeter wave source development efforts at UCLA, *USA-JAPAN Workshop on Submillimeter Diagnostic Techniques,* pp. 170–205.

Bozler, C. O., and Alley, G. D., (1980). Fabrication and numerical simulation of the permeable base transistor, *IEEE Trans. Electron Devices,* Vol. **ED-27,** pp. 1128–1141.

Chang, K., Kung, J. K., Asher, P. G., Hayashibara, G. M., and Ying, R. S., (1981). GaAs Read-type IMPATT diode for 130 GHz CW operation *Electron. Lett.* Vol. **17,** pp. 471–473.

Chive, M., Constant, E., Lefebvre, M., and Pribetich, J., (1975). Effect of tunneling on high-efficiency IMPATT avalanche diodes, *Proc. IEEE (Lett.),* Vol. **63,** pp. 824–826.

Elta, M. E., and Haddad, G. I., (1978). Mixed tunneling and avalanche mechanisms in *p-n* junctions and their effects on microwave transit-time devices, *IEEE Trans. Electron Devices,* Vol. **ED-25,** pp. 694–702.

Elta, M. E., and Haddad, G. I., (1979). Large-signal performance of microwave transit-time devices in mixed tunneling and avalanche breakdown, *IEEE Trans. Electron Devices,* Vol. *ED*-27, pp. 941–948.

Elta, M. E., and Haddad, G. I., (1979). High-frequency limitations of IMPATT, MITATT and TUNNETT mode devices, *IEEE Trans. Microwave Theory Tech.,* Vol. **MTT-27,** pp. 442–449.

Elta, M. E., Fetterman, H. R., Macropoulos, W. V., and Lambert, J. J., (1980). 150 GHz GaAs MITATT source, *IEEE Electron Device Lett.* Vol. **EDL-1,** pp. 115–116.

Ema, Y., and Motoya, K., (1985), Contact interface between the metal and compound semiconductor, *The report to the Nippon Gakujutsu Sinko-kai* (in Japanese).

Gibbons, G., Purcell, J. J., Wickens, P. R., and Gokgor, H. S., (1972). 50 GHz gallium arsenide IMPATT oscillator, *Electron. Lett.,* Vol. **8,** 513–514.

Gunn, J. B., (1956). Avalanche injection in semiconductors, *Proc. Phys. Soc. London Ser. B.,* Vol. **69,** pp. 781–790.

Gupta, M. S., (1975). A simple approximation method of estimating the effect of carrier diffusion in IMPATT diodes, *Solid-State Electron.,* Vol. **18,** 327–330.

Ishibashi, T., Nakamura, T., and Ohmori, M., (1977). Submillimeter wave silicon impatt diodes, *IECE Jpn. Tech. Group Meeting* **MW76-137** (in Japanese).

Ishihara, O., Nishitani, K., Sawano, H. and Mitsui, M., (1976). Ohmic contacts to p-type GaAs, *Jpn. J. Appl. Phys.,* Vol. **15,** pp. 1411–1412.

Kuno, H. J., (1979). IMPATT devices for generation of millimeter waves, *Infrared and Millimeter Waves,* Vol. **1,** *Source of Radiation,* Chapter 2, Academic Press, New York.

Kuvås, R., Lee, C. A., (1970). Carrier diffusion in semiconductor avalanches, *J. Appl. Phys.*, Vol. **41,** pp. 3108–3116.

Kwok, S. P., and Haddad, G. I., (1972). Effects of tunneling on an IMPATT oscillator, *J. Appl. Phys.*, Vol. **43,** pp. 2830–3830.

McKay, K. G., and McAfee, K. B., (1953). Electron multiplication in silicon and germanium, *Phys. Rev.* Vol. **91,** pp. 1079–1084.

Misawa, T., (1966). Negative resistance on *p-n* junction under avalanche breakdown conditions, Part I and II, *IEEE Trans. Electron Devices,* Vol. **ED-13,** pp. 137–151.

Misawa, T., (1967). Multiple uniform layer approximation in analysis of negative resistance in *p-n* junctions in breakdown, *IEEE Trans. Electron Devices,* Vol. **ED-16,** pp. 64–77.

Motoya, K., and Nishizawa, J., (1985). TUNNETT, *Int. J. Infrared and Millimeter Waves,* Vol. **6,** pp. 483–495.

Nawata, K., Ikeda, M., and Ishii, Y., (1974). Millimeter-wave GaAs Schottky-barrier IMPATT diodes, *IEEE Trans. Electron Devices,* Vol. **ED-21,** pp. 128–130.

Nishitani, K., Ishihara, O., Sawano, H., Mitsui, S., and Shirahata, K., (1975). Characteristics of pulsed GaAs IMPATT diodes, *IECE of Japan, Technical Group Meeting,* **SSD75-33** (in Japanese).

Nishizawa, J., (1963). Future of semiconductor laser, *Denshi Kagaku,* Vol. **13**(4), pp. 17–20, 30–31 (in Japanese).

Nishizawa, J., (1971). Panel discussion of several problems on semiconductor oscillators, *Joint Conv. Record Four Inst. Elec. Eng., Jpn.* pp. 121–144.

Nishizawa, J., (1974). Progress of compound semiconductor devices, *Denshi-Zairyo,* pp. 18–22 (in Japanese).

Nishizawa, J., (1978). Panel discussion on SIT, *Semicond. Electron.* (*Handotai-Kenkyu*) Vol. **15,** pp. 319–358.

Nishizawa, J. (1979). Recent progress and potential of SIT, *Digest Tech. Papers, Int. Conf. Solid State Devices,* **11th,** August A-0-1.

Nishizawa, J., (1980). Recent progress and potential of SIT, *Jpn. J. Appl. Phys. Suppl.* **19-1,** Vol. **19,** pp. 3–11.

Nishizawa, J., and Suto, K., (1980). Semiconductor Raman laser, *J. Appl. Phys.* Vol. **51,** pp. 2429–2431.

Nishizawa, J., and Watanabe, Y., (1953), *The Contract Research Report to the Nippon Telegraph and Telephone Public Corporation.*

Nishizawa, J., and Watanabe, Y., (1958). High frequency properties of the avalanching negative resistance diode, *Sci. Rep. Res. Inst. Tohoku Univ.* Vol. **10**(2), pp. 91–108.

Nishizawa, J., and Yamamoto, K., (1978). High frequency high power static induction transistor, *IEEE Trans. Electron Devices,* Vol. **ED-25,** pp. 314–322.

Nishizawa, J., Ohmi, T., and Sakai, T., (1974). Millimeter-wave oscillation from TUNNETT diode, *Proc. Eur. Microwave Conf.* pp. 449–453.

Nishizawa, J., Terasaki, T., and Shibata, J., (1975a). Field-effect transistor versus analog transistor (static induction transistor), *IEEE Trans. Electron Devices,* Vol. **ED-22,** pp. 185–197.

Nishizawa, J., Okuno, Y., and Tadano, H., (1975b). Nearly perfect crystal growth of III-V compounds by the temperature difference method under controlled vapor pressure, *J. Crystal Growth,* Vol. **31,** 215–222.

Nishizawa, J., Suto, K., and Techima, T., (1977a). Minority-carrier lifetime measurements of efficient GaAlAs *p-n* heterojunctions, *J. Appl. Phys.* Vol. **48,** pp. 3484–3495.

Nishizawa, J., Motoya, K., and Okuno, Y., (1977b). 200 GHz TUNNETT diodes, *Proc. Conf. Solid State Devices,* **9th** August C-2-2.

Nishizawa, J., Motoya, K., and Okuno, Y., (1978a). GaAs TUNNETT diodes, *IEEE Trans. Microwave Theory Tech.* Vol. **MTT-26,** pp. 1029–1035.

Nishizawa, J., Motoya, K., and Okuno, Y., (1978b). *Jpn. J. Appl. Phys. Suppl.* **17-1,** Vol. **17,** pp. 167–172.

Nishizawa, J., Ohmi, T., and Niranjian, M. S., (1978c). Avalanche induced dispersion in Impatt diodes, *Solid-State Electron.* Vol. **21,** pp. 847–858.

Nishizawa, J., Motoya, K., and Okuno, Y., (1979). Submillimeter wave oscillation from GaAs TUNNETT diodes, *Proc. Eur. Microwave Conf.,* **4th,** pp. 463–467.

Nishizawa, J., Motoya, K., and Suzuki, K., (1980), GaAs p^+-n^+-$i(\nu)$-n^+ tunnett diodes, *Proc. Eur. Microwave Conf.,* **10th,** pp. 667–671.

Nishizawa, J., (1975). TUNNETT diode, *Oyo Butsuri,* Vol. **44,** pp. 821–825 (in Japanese).

Nishizawa, J., (1982). The GaAs TUNNETT diodes, *Infrared and Millimeter Waves* (ed. by K. J. Button), Vol. **5,** Chapter 4, pp. 215–266, Academic Press, New York.

Nishizawa, J., and Suto, K., (1983). Semiconductor Raman and Brillouin lasers for far-infrared generation, *Infrared and Millimeter Waves* (ed. by K. J. Button), Vol. **7,** Chapter 6, Academic Press, New York.

Nishizawa, J., Abe, H., and Kurabayashi, T., (1984). Photo Molecular Layer Epitaxy of GaAs, *Tech. Group Meeting of IECE of Japan,* **SSD 84-55** (in Japanese).

Nishizawa, J., and Kokubun, Y., (1984). Recent progress in low temperature photochemical processes, *Extended Abstracts of 16th Int. Conf. on Solid State Devices and Materials,* Kobe, pp. 1–4.

Nishizawa, J., Abe, H., and Kurabayashi, T., (1985). Molecular Layer Epitaxy, *J. Electrochem. Soc.* Vol. **132,** pp. 1197–1200.

Nishizawa, J., and Motoya, K., (1984). TUNNETT, *9th Int. Conf. on Infrared and Millimeter Waves,* **W-1-2,** pp. 223–226.

Nishizawa, J., Abe, H., and Kurabayashi, T., (1985). Molecular Layer Epitaxy, *J. Electrochem. Soc.* Vol. **132,** pp. 1197–1200.

Ohmi, T., and Motoya, K., (1976). Millimeter wave oscillations from TUNNETT diodes, *IECE Jpn. Tech. Group Meeting,* Vol. **ED-75-71** (in Japanese).

Okabe, T., and Nishizawa, J., (1969). Some considerations of tunnel injection transit-time (TUNNETT) diode oscillator, *IECE Jpn. Tech. Group Meeting,* Vol. **ED-69-19.**

Okabe, T., Takamiya, S., Okamoto, K., and Nishizawa, J., (1968). Bulk oscillation by tunnel injection, presented at IEEE Int. Electron Devices Meeting, December; also, *RIEC Technical Rep.* **TR-31,** Tohoku Univ.

Okamoto, K., Nishizawa, J., and Takahashi, K., (1965). Measurement of hot carrier diffusion constant in semiconductors, *J. Appl. Phys.* Vol. **36,** pp. 3716–3722.

Pan, D. S., and Lee, N., (1981). GaAs abrupt MITATT and TUNNETT, *6th Int. Conf. on Infrared and Millimeter Waves,* **M-5-3.**

Pan, D. S., and Lee, N., (1981). Narrow band-gap semiconductors for TUNNETTs, *6th Int. Conf. on Infrared and Millimeter Waves,* **M-5-8.**

Pidgeon, C. R., Lax, B., Aggarwal, R. L., and Chase, C. E., (1971). Tunable coherent radiation source in the 5μ region, *Appl. Phys. Lett.,* Vol. **19.**

Read, W. T., (1958). A proposed high-frequency negative resistance diode, *B.S.T.J.,* Vol. **37,** pp. 401–466.

Schwarz, R. I., and Bonek, E., (1978). Current tuned GaAs Schottky-barrier IMPATT diodes for 60-96 GHz operation, *Electron. Lett.,* Vol. **14,** pp. 812–814.

Schwartz, R., Thim, H. W., Potzl, H. W., (1977). Influence of diffusion on the small-signal properties of Misawa diodes, *Electron. Lett.,* Vol. **13,** pp. 288–289.

Semichon, A., Michel, J., Constant, E., and Vanoverschelde, A., (1970). Microwave oscillation of a tunnel transit-time diode, in *Proc. 8th Int. Conf. on Microwaves and Optical Generation and Amplification* (Amsterdam, The Netherlands), pp. 7.15–7.20.

Shimizu, A., and Nishizawa, J., (1961). Alloy-diffused variable capacitance diode with large figure-of-merit, *I.R.E. Trans. Electron Devices,* Vol. **ED-8,** pp. 370–377.

Shockley, W., (1954). Negative resistance arising from transit time in semiconductor diodes, *Bell Syst. Tech. J.*, Vol. **33,** p. 799.

Shur, M. S., and Eastman, L. F., (1979). Ballistic transport in semiconductor at low temperature for low-power high-speed logic, *IEEE Trans. Electron Devices,* Vol. **ED-26,** pp. 1677–1683.

Suto, K., and Nishizawa, J., (1984). Semiconductor Raman laser, *9th Int. Conf. on Infrared and Millimeter Waves,* **M-7-5,** pp. 69–70.

Sze, S. M., and Gibsons, G., (1966). Avalanche breakdown voltages of abrupt and linearly graded *p-n* junctions in Ge, Si, GaAs and GaP, *Appl. Phys. Lett.,* Vol. **8,** p. 111.

Tyagi, M. S., (1968). Zener and avalanche breakdown in silicon alloyed *p-n* junctions, *Solid-State Electron.,* Vol. **11,** pp. 95–115.

Watanabe, Y., and Nishizawa, J., (1952). Reverse characteristic of the semiconductor rectifier, *Record Elec. Commun. Eng. Conversazione, Tohoku Univ.* Vol. **21**(3).

Watanabe, Y., Nishizawa, J., Yamamoto, T., and Shimizu, A., (1960). Wide band parametric amplification of the distributed semiconductor diode, *IECE Jpn., Tech. Committee Meeting,* January.

CHAPTER 2

Computer-Aided Testing of SIS Junctions and Solid-State Devices

Martin V. Schneider

AT&T Bell Laboratories
Crawford Hill Laboratory
Holmdel, New Jersey 07733

Denise A. Brown

NYNEX Information Resources
Lynn, Massachusetts 01901

I. Introduction

Solid-state devices which are used as sources, amplifiers, modulators and detectors in microwave and optoelectronic circuits display nonlinear current-voltage characteristics. The nonlinearity is a major factor affecting the performance of the device in a specific circuit. A typical example is an SIS (superconductor-insulator-superconductor) junction whose mixing properties are determined by the ski-sloped shape of the current-voltage curve. Other examples are frequency converter diodes, whose first and second order derivatives of current versus voltage affect their noise temperature and conversion loss. A third device which displays

ISBN 0-12-147700-2

nonlinearity is an FET, whose transfer characteristic determines the transconductance of the transistor. For many applications it is sufficient to test the device characteristic with an oscilloscope. To perform accurate tests, however, and to compare the properties of different batches, it becomes necessary to conduct measurements with a computer-controlled current or voltage source and to display the data with a graphic plotter.

The purpose of this paper is to describe the instrumentation and software needed to perform both ambient and temperature-variable measurements on a number of two-terminal solid-state devices. The programs can be readily adapted to three-terminal devices or to diodes which require testing of the capacitance versus voltage characteristic, such as varactor diodes and photodetectors. Two specific examples which are discussed in this paper are automated tests performed on SIS junctions and the software for measuring temperature-variable characteristics of Schottky barrier diodes. The SIS junctions are devices which are needed in microwave and millimeter-wave front ends requiring the attainment of quantum-limited noise performance. The Schottky diodes are widely used in frequency converters, multipliers, modulators and switching circuits. Both devices are characterized by conductance and other parameters which can be computed from the test data using appropriate algorithms. The performance of a device can be predicted from these parameters and some other properties, such as the device capacitance and its parasitics in a specific circuit.

II. Nonlinearities in SIS Junctions, Schottky Diodes and FETS

The three basic nonlinear current-voltage characteristics of devices which are used for detectors, frequency converters and amplifiers are shown schematically in Figs. 1–3. Fig. 1 displays the characteristic of an SIS tunnel junction of the Pb alloy type which has been used in 85-115 GHz front ends by Woody, Miller and Wengler (1985). The ski-sloped curve consists of three distinctive parts: a low-field leaky conductance G_L below the bandgap, a transition region near V_G, and an ohmic part with conductance G_N above the gap. The heterodyne performance of the device can be calculated from this curve using the quantum mechanical theory of mixing developed by Tucker (1979). In addition to the ski-sloped curve caused by single particle tunneling, a fixed current is observed at zero bias which arises from the tunneling of superconducting pairs (Josephson current). Both the pair and single particle current contributions can be displayed on the same graph if a programmable current source is used for the CAM test procedure.

The nonlinear characteristic of a metal-semiconductor junction (Schot-

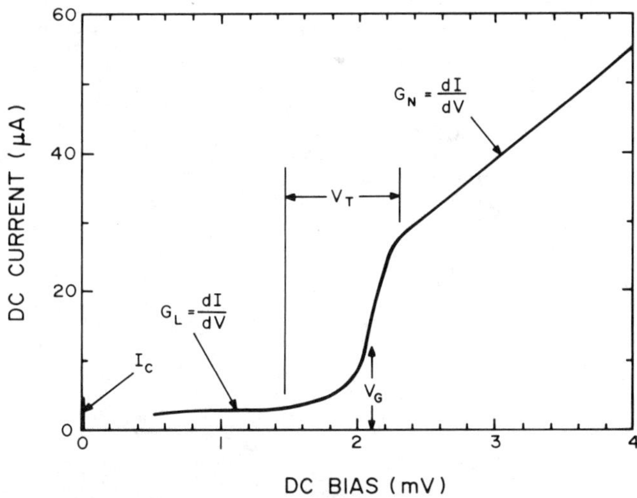

FIG. 1 Nonlinear current-voltage characteristics of a typical superconductor-insulator-superconductor junction (SIS junction) developed for a 100 GHz low-noise receiver by Woody, Miller and Wengler (1985). G_L is the low field leaky conductance below the bandgap and G_N the ohmic conductance above the bandgap transition near V_G. I_C is the Josephson supercurrent and V_T the transition width of the PbInAu junction.

tky barrier) is shown in Fig. 2. The Pt-GaAs junction fabricated by Verlangieri and Schneider (1985) has an *i-v* characteristic given by (Rhoderick, 1978)

$$i = I_s \exp\left[\frac{q(v - iR)}{\eta kT}\right] \tag{1}$$

where I_s is the saturation current, η the ideality factor, R the series resistance and T the device temperature. The electron charge is $q = 1.602 \times 10^{-19}$ Coulomb and the Boltzmann constant is $k = 1.381 \times 10^{-23}$ Joule/K. An algorithm for calculating the device parameters I_s, R and η from a sequence of test data can be derived from Eq. (1) by generalization of a numerical procedure reported by Lidholm (1977). This algorithm is derived in Appendix A. The device parameters R and η determine the equivalent noise temperature of the diode in a frequency converter circuit. For example, if the major source of noise is shot noise (Viola and Mattauch, 1973), then the equivalent noise temperature of the diode, T_{eq}, is

$$T_{eq} = \frac{\eta}{2} T. \tag{2}$$

The equations for the conversion loss as a function of the device parameters, and for the noise temperature with $R \neq 0$ are given in recent papers

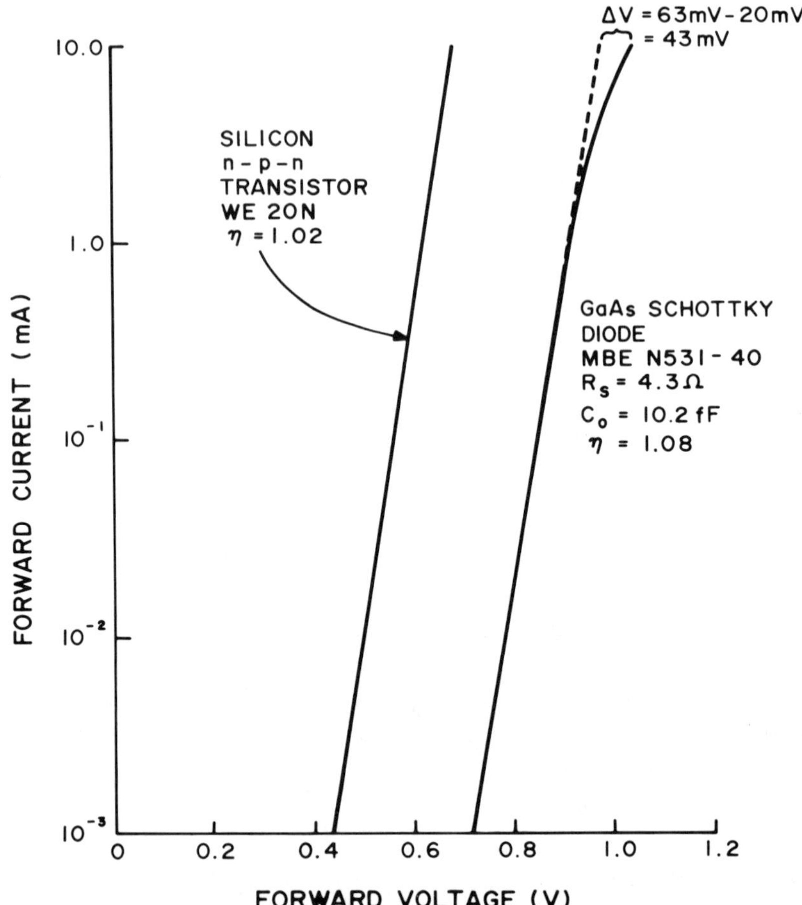

FIG. 2 Forward current-voltage characteristic of a millimeter-wave metal-semiconductor microjunction (Verlangieri and Schneider, 1985). The voltage drop Δv is caused by the series resistance in the junction. The characteristic of a silicon transistor is shown as a reference.

by Jelenski et al. (1984), Crowe and Mattauch (1985) and Schneider (1982).

Fig. 3 shows the transfer characteristic of an FET. The drain current, i_D, is a quadratic function of the gate-source voltage, v_{GS}, and is given by (Sze, 1981)

$$i_D = I_{DSS}\left(1 - \frac{v_{GS}}{v_P}\right)^2 \tag{3}$$

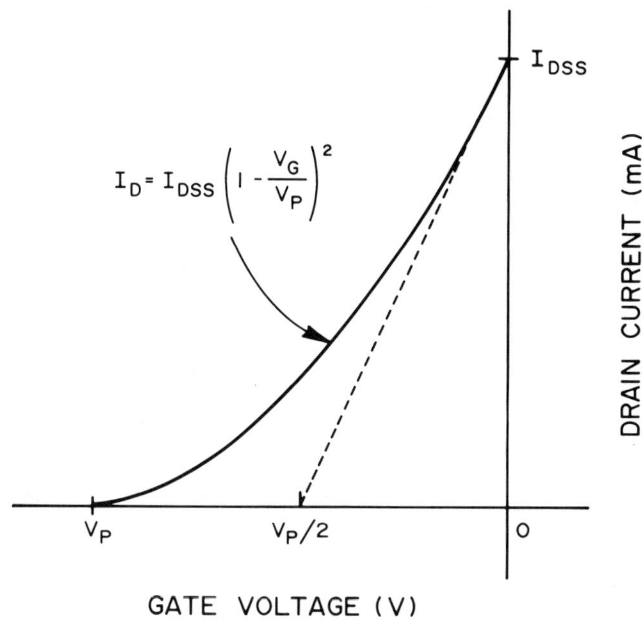

FIG. 3 Transfer characteristic of FET showing the drain current I_D as a function of the gate voltage V_G. I_{DSS} is the drain saturation current and V_P the pinch-off voltage of the FET.

where I_{DSS} is the drain saturation current and v_P the pinch-off voltage. From Eq. (3) one obtains the transconductance, g_m

$$g_m = \frac{\delta i_D}{\delta v_{GS}} = \frac{2I_{DSS}}{v_P}\left(1 - \frac{v_{GS}}{v_P}\right). \tag{4}$$

As discussed by Fukui (1979), the magnitude of the transconductance of a good device can be assumed to remain constant up to nearly the cutoff frequency. Precise methods for determining other relevant MESFET parameters such as the pinch-off voltage, active channel resistance and effective gate length from non-linear device characteristics have also been treated by Fukui (1979).

The computer program and test apparatus for generating graphs of the various nonlinear device characteristics are described in subsequent sections. The software also provides a listing of the relevant device parameters for the case of metal-semiconductor junctions. These parameters are printed for a fixed temperature or variable temperatures ranging from 15 K to 300 K.

III. Description of Test System

The test system for measuring the current-voltage characteristic is shown schematically in Fig. 4. The system consists of a network for connecting and protecting the device under test, a number of input–output devices and a controller. Instrument control and data transfer are enabled through the controller, which is interfaced with the input–output devices on the IEEE-488 BUS. A similar test system for performing automated tests on power amplifiers has been described by Carlsen (1983).

A dc current which is generated by a Keithley Model 220 current source is delivered to the device under test. The currents are produced over a user specified range from 1 nA to 100 mA with an accuracy of 0.1% and a minimum step size of 0.5 nA. The corresponding voltages are measured with either two HP 3456A digital voltmeters or two HP 3478A multimeters, depending on whether accuracy or speed is desired. The resolution of the HP 3456A is 100 nanovolts at a test rate of 48 readings/second. This gives an accuracy of $\pm 0.002\%$ for diodes with knee voltages of the order of 0.5 Volt. The knee voltage is defined as the voltage at which the device carries 1 μA of forward current. The HP 3478A has an accuracy of 0.005%, but compensates for this by performing measurements at higher rates.

The temperature of the device under test is monitored by a Lakeshore digital cryogenic thermometer. Temperature changes are detected by a silicon diode sensor which is connected to the thermometer. A closed cycle helium refrigerator cools the device under test from 300 K to 15 K. Execution of the program results in a complete measurement of the current voltage characteristics at integer multiples of 10 K during the cooling cycle, e.g., at 290 K, 280 K down to 20 K. A final test is performed after reaching a device temperature of 15 K. The thermometer measures the temperature changes with a resolution of 0.01 K below 30 K, 0.05 K between 30 K and 100 K, and 0.1 K above 100 K.

IV. Software

The system software is written in BASIC for a Hewlett Packard Model 3456A desktop computer. The program is divided into three sections, described below. These are: program setup and initialization, testing sequence and printing, and plotting of the relevant measurement data. The system provides user options with menu driven software. It is tutorial in nature and begins with a description of the task to be performed and specifies the instruments to be used. The user selects the measuring devices from the menu. The user is then asked to provide his or her initials, the date and the device number.

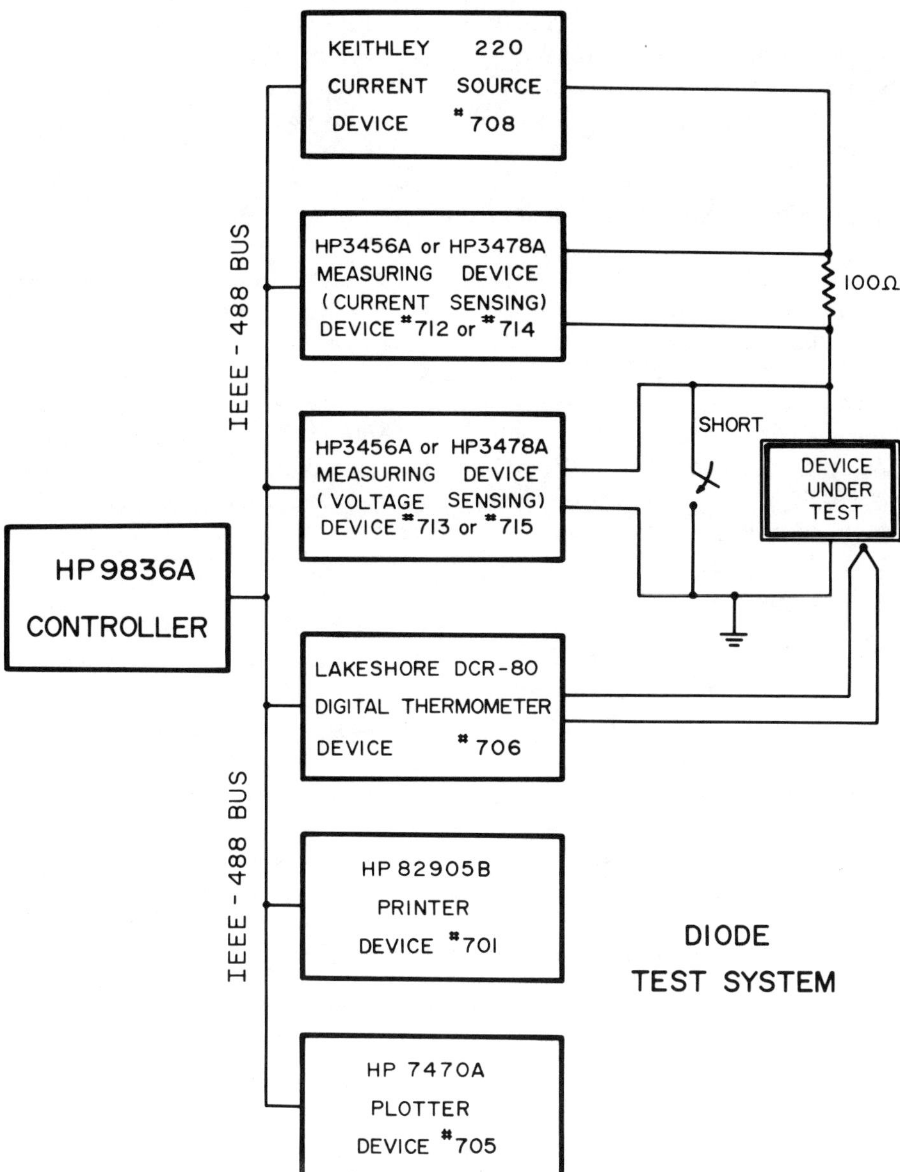

FIG. 4 Schematic view of automated test apparatus for measuring the temperature-variable current-voltage characteristic of solid-state devices. Instrument control and data transfer are enabled through the controller, which is interfaced with the input/output devices on the IEEE-488 BUS.

A prompt then requests a choice of one of three testing sessions. In the case that any major mistake is made, the program can be reset and restarted. Memory space is then allocated for the data and specified currents to be produced by the current source. This section ends with a pause in the program before the actual measurements begin.

The measurement is first made by measuring the temperature and subsequently increasing the current from 1 μA to 3 mA. At each current the voltage is measured simultaneously across the device and the fixed 100 Ohm precision resistor shown in Fig. 4. The currents are programmed to increase in the sequence $k \cdot 10^n$ μA with $k = 1, 2, 3, 6$ and $n = 0, 1, 2, 3$ up to a maximum current of 3 mA. It is to be noted that the programmable current source shown in Fig. 4 has an accuracy of 0.05% $\pm$ 10 nA for the lowest required range of 10 μA. Thus the actual current flowing through the precision resistor and the device will only approximate the sequence of desired currents. In order to find the voltage across the device for a desired current such as for example 2.000 μA, the program will automatically set two currents which bracket the desired current and interpolate the corresponding voltages to obtain the voltage for the desired current of 2.000 μA.

After reaching the maximum current a second temperature measurement is made followed by a measurement which produces a hysteresis curve. The second measurement is identical to the first except that the currents are generated in reverse order. This cycle ends with a third temperature reading. The measurements are then used to calculate the series resistance, ideality factors, saturation current and expovoltage $V_0 = \eta kT/q$. The printer displays the data and calculated values. In addition to these printed displays, a curve of the current-voltage characteristic is plotted at 300 K, 200 K, 100 K, and 15 K.

The final measurement sequence occurs at 15 K. The values of $V_0 = \eta kT/q$ and the mid-test temperature for each cycle are printed and plotted to complete the automated test. The temperature variable tests can run unattended overnight. If necessary, the software and instrumentation can be readily expanded to test a large number of devices in parallel or in sequence.

V. Measurements

The current-voltage characteristics measured for two different devices are displayed in Figs. 5 and 6. The symmetrical characteristic shown in Fig. 5 was obtained for an SIS junction which was made for use in a 230 GHz low-noise front end. The geometry and fabrication of this device are described in a recent paper by Woody, Miller and Wengler (1985). The

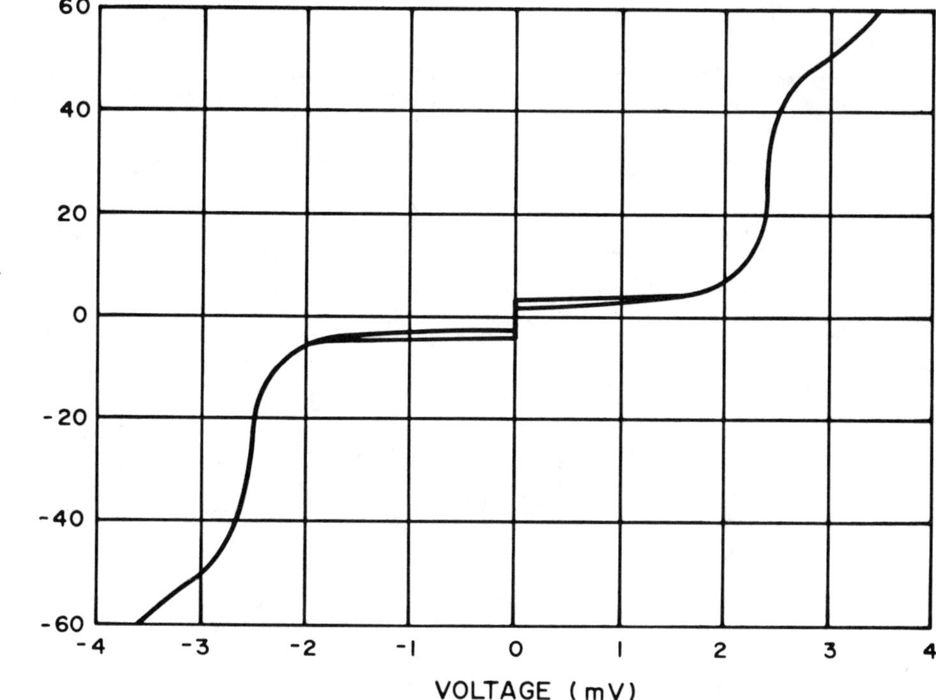

FIG. 5 Symmetrical current-voltage plot obtained for SIS junction at liquid helium temperature. Nearly ideal switching is achieved at 2.4 mV which is equal to the band gap of the Pb alloy junction (Woody, Miller and Wengler, 1985).

measurements were performed with both the test system shown in Fig. 4 and a programmable Hewlett-Packard Model 4145A semiconductor parameter analyzer. A low-noise 40 dB preamplifier was needed in front of the analyzer in order to improve its voltage sensitivity from three to five significant digits. The preamplifier improved the accuracy of the voltage measurement across the device from ± 100 μV to ± 1 μV, which is sufficient for junctions with an energy gap of about 2 eV. The results obtained for a specific SIS junction with the semiconductor analyzer are shown in Fig. 5. The three regions showing a fixed current below the bandgap, a rapid current increase in the vicinity of the gap and the breaking up of Cooper pairs at higher voltages are shown in the Figure.

The temperature-variable current-voltage characteristics obtained for a metal-semiconductor junction at 15 K, 100 K, 200 K and 295 K are displayed in Fig. 6. The tests were performed with the automated system of Fig. 4, using a CTI Model 350 SC closed cycle helium refrigerator. The

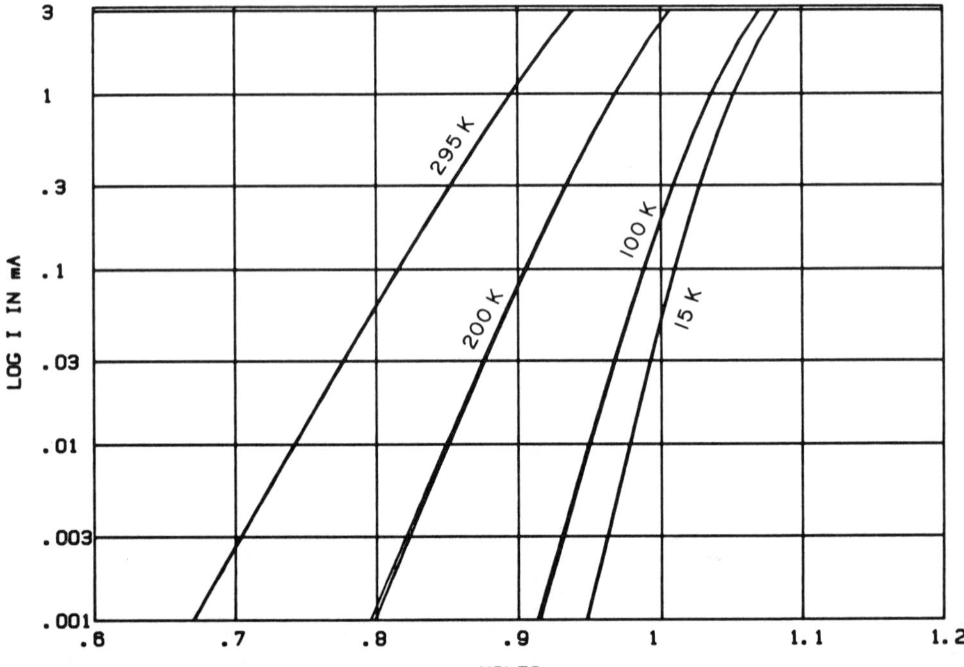

FIG. 6 Current-voltage characteristic for GaAs Schottky barrier measured at 15 K, 100 K, 200 K and 295 K. The characteristic is an exponential function at all temperatures with a small deviation of high currents caused by the series resistance of the device.

junction, whose geometry and fabrication was described by Verlangieri and Schneider (1985), consists of an epitaxial GaAs substrate with a 2 μm diameter Pt anode surrounded by an insulating SiO_2 layer. The GaAs substrate was doped with Si at $1 \cdot 10^{18}$ cm^{-3}, and the epitaxial layer with a thickness of 700 Å had a Si doping concentration of $n = 1 \cdot 10^{16}$ cm^{-3}. The current-voltage characteristic is exponential at all temperatures, with a small deviation at high currents caused by the series resistance in the undepleted epitaxial layer and the spreading resistance in the GaAs substrate. The small apparent hysteresis at the test temperature of 200 K is caused by the fact that the device temperature decreases by 2 K between the beginning and the end of the test cycle. This effect becomes less pronounced at lower temperatures since the device characteristic does

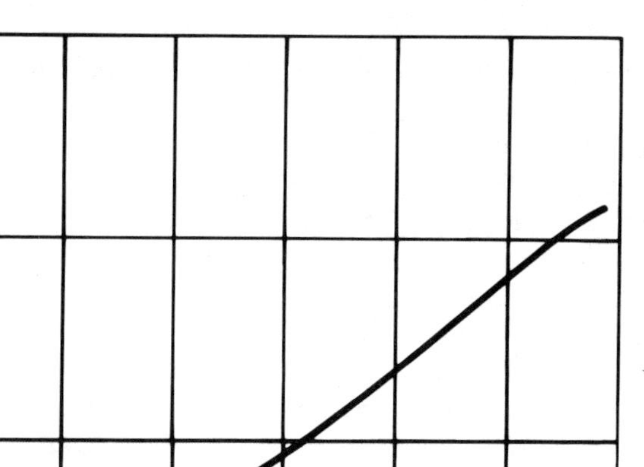
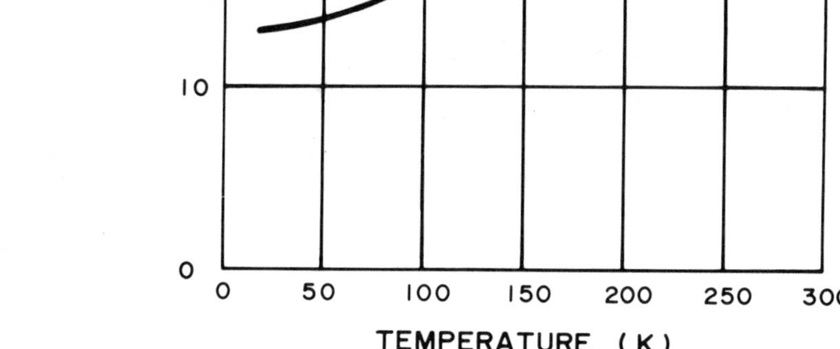

FIG. 7 Plot of $V_0 = \eta kT/q$ as a function of device temperature. In the linear portion above 100 K electrons cross the barrier at the interface of the junction by thermionic emission. At temperatures below 100 K electron transport occurs mainly by tunneling which is not temperature dependent.

not change rapidly with decreasing temperature. Fig. 7 shows a plot of $V_0 = \eta kT/q$ as a function of the device temperature. In the linear portion above 100 K electrons cross the barrier at the interface of the junction by thermionic emission. At temperatures below 100 K electron transport occurs by tunneling which is not temperature dependent.

The algorithms for calculating the series resistance R, ideality factor η and saturation current I_s are derived in Appendix A. Numerical methods which allow an accurate determination of the series resistance R have been also treated by Jelenski et al. (1985). A printout of the current-voltage data obtained for a GaAs Schottky diode (Verlangieri and Schneider, 1985) is shown in Appendix B. The measurement was performed at a device temperature of 200 K for forward currents ranging from 1 μA to 3 mA. The ideality factor for increasing currents is $n = 1.362$ and for decreasing currents $H(n) = 1.338$. At ambient room temperature both factors are about 1.12. The small difference between n and $H(n)$ and the hysteresis in the current-voltage data are caused by the temperature change from 201.6 K to 199.3 K between the beginning and the end of the test cycle. Appendix C is a listing of the program and Appendix D shows the flowchart for performing the measurements and calculating the device parameters. It is assumed in these calculations that the device can be characterized by a single exponential term, i.e. that leakage currents and variations of the barrier height caused by multiple chemical phases or microclusters can be neglected. This assumption is valid if the device is fabricated using current state-of-the-art techniques.

VI. Conclusions

The software and computer-controlled instrumentation described in this paper are useful for measuring and characterizing the nonlinear properties of solid-state devices. The testing can be performed at fixed or variable temperatures ranging from 15 K to 300 K. The system has been used successfully for testing both SIS devices and frequency converter diodes.

References

Carlsen, E. R. (1982). Automated measurement system for characterizing power amplifier performance, *1983 IEEE MTT-S International Microwave Symposium Digest,* pp. 381–383.

Crowe, T. W., and Mattauch, R. J. (1985). Conversion losses in room temperature and cryogenic schottky barrier mixer diodes, *IEEE Trans. Microwave Theory and Techniques,* vol. **MTT-33,** to be published.

Fukui, H. (1979). Determination of the basic device parameters of a GaAs MESFET, *B.S.T.J.,* vol. **58,** pp. 771–797.

Jelenski, A., Kollberg, E., Schneider, M. V., and Zirath, H. (1985). Accurate determination of the series resistance of MM-wave Schottky diodes, *Proc. 15th European Microwave Conference,* pp. 279–284.

Jelenski, A., Schneider, M. V., Cho, A. Y., Kollberg, E. R., and Zirath, H. (1984). Noise measurements and noise mechanisms in microwave mixer diodes, *1984 IEEE MTT-S International Microwave Symposium Digest,* pp. 552–554.

Lidholm, S. (1977). Low-noise mixers for 80-120 GHz, *Research Report No. **129,** Research Laboratory of Electronics and Onsala Space Observatory, Chalmers University of Technology,* Goteborg, Sweden.

Rhoderick, E. H. (1978). Metal-Semiconductor Contacts. Oxford Univ. Press (Clarendon), London and New York.

Schneider, M. V. (1982). Metal-semiconductor junctions as frequency converters. *Infrared and Millimeter Waves*, vol. **6,** edited by K. F. Button, pp. 210–275. Academic Press, New York.

Sze, S. M. (1981). *Physics of Semiconductor Devices*. Wiley, New York.

Tucker, J. R. (1979). Quantum limited detection in tunnel junction mixers, *IEEE Journal of Quantum Electronics*, vol. **QE-15,** pp. 1234–1258.

Verlangieri, P. A., and Schneider, M. V. (1985). Microfabrication of GaAs Schottky diodes for multipliers, mixers and modulators, *International Journal of Infrared and Millimeter Waves*, vol. **6,** pp. 1191–1201.

Viola, T. J., Jr., and Mattauch, R. J. (1973). Unified theory of high-frequency noise in Schottky barriers, *J. Appl. Phys.*, vol. **44,** pp. 2805–2808.

Woody, D. P., Miller, R. E., and Wengler, M. F. (1985). "85 to 115 GHz Receivers for Radio Astronomy," *IEEE Trans. Microwave Theory and Techniques*, vol. **MTT-33,** pp. 85–89.

Appendix A
Calculation of Series Resistance Ideality Factor and Saturation Current

Let us assume that the current-voltage characteristic of a nonlinear device is given by Eq. (1)

$$i = I_s \exp\left(\frac{qu}{\eta kT}\right) \tag{A-1}$$

where u is the voltage across the junction given by $u = v - Ri$. If we conduct a sequence of measurements with $n = 1, 2 \ldots$ we obtain

$$u_n = v_n - Ri_n \tag{A-2}$$

$$\ln i_n = \frac{qu_n}{\eta kT} + \ln I_s. \tag{A-3}$$

The parameters R, I_s and η can be conveniently found by introducing the differences

$$\Delta v_n = v_{n+1} - v_n \tag{A-4}$$

$$\Delta v_m = v_{m+1} - v_m \tag{A-5}$$

$$\Delta \log i_n = \log\left(\frac{i_{n+1}}{i_n}\right) \tag{A-6}$$

$$\Delta \log i_m = \log\left(\frac{i_{m+1}}{i_m}\right) \tag{A-7}$$

From Eqs. (A-2) to (A-7) one obtains

$$R_{nm} = \frac{\Delta v_n \cdot \Delta \log i_m - \Delta v_m \cdot \Delta \log i_m}{\Delta i_n \cdot \Delta \log i_m - \Delta i_m \cdot \Delta \log i_n}. \tag{A-8}$$

Equation (A-8) can be simplified substantially if one chooses i_n and i_m such that $\Delta \log i_n = \Delta \log i_m = 1$ resulting in

$$R_{nm} = \frac{\Delta v_n - \Delta v_m}{\Delta i_n - \Delta i_m}. \tag{A-9}$$

For the special case $i_m = 10\ \mu\text{A}$, $i_{m+1} = 100\ \mu\text{A}$, $i_n = 1$ mA and $i_{n+1} = 10$ mA one obtains $\Delta i_m = 0.09$ mA and $\Delta i_n = 9$ mA. This gives for R_{nm}

$$R_{nm} = \frac{\Delta v_n - \Delta v_m}{8.91}\ \text{Ohm} \tag{A-10}$$

if the voltages are measured in mV. It is to be noted that $R_{nm} \equiv R$ if the characteristic is an ideal exponential and that the labeling of currents and voltages can be arbitrary.

The expovoltage $V_0 = \eta kT/q$ defined in Section IV and the saturation current I_s are obtained from Eqs. (A-2) to (A-7) and are given by

$$V_0 = \frac{\Delta i_n \cdot \Delta v_m - \Delta i_m \cdot \Delta v_n}{\Delta i_n \cdot \Delta \log i_m - \Delta i_m \cdot \Delta \log i_n} \tag{A-11}$$

$$ln\ I_s = \frac{Ri_n - v_n}{V_0} + ln\ i_n. \tag{A-12}$$

In the program listed in Appendix C only three currents are used to calculate R, V_0, and I_s. The currents are $i_m = 10\ \mu\text{A}$, $i_{m+1} = i_n = 100\ \mu\text{A}$, and $i_{n+1} = 1$ mA. Equation (A-10) thus becomes $R = (\Delta v_n - \Delta v_n)/0.81$ Ohm if the voltages are measured in mV.

Appendix B
Typical Output for Diode Test Performed for a Device Temperature of 200 K and a Forward Current from 1 μA to 3 mA

TEST DATE:
DIODE NUMBER: 17H61-31
MEASURING DEVICE: HP3456A-accuracy
OPERATOR: CT

POINT RESISTANCE = 1.50 OHMS

CYCLE# 15

Initial Temp. is 201.6 K

Current in mA	Volts	Hysteresis
1.0000E-06	0.79569	0.79861
2.0000E-06	0.81159	0.81419
3.0000E-06	0.82097	0.82331
6.0000E-06	0.83708	0.83913
1.0000E-05	0.84907	0.85088
2.0000E-05	0.86548	0.86701
3.0000E-05	0.87522	0.87656
6.0000E-05	0.89205	0.89316
1.0000E-04	0.90473	0.90568
2.0000E-04	0.92243	0.92316
3.0000E-04	0.93323	0.93377
6.0000E-04	0.95251	0.95292
1.0000E-03	0.96778	0.96807
2.0000E-03	0.99087	0.99103
3.0000E-03	1.00633	1.00637

Temp During Test is 200.1 K

R = 10.14 OHMS
n = 1.362
$H(n)$ = 1.338
I_0 = 2.123E-18 mA
V_0 = 23.53 mV

Final Temp is 199.3 K

Appendix C
Program Listing

```
10     !This is the Diode testing Program AUTO_TEST
20     !The automatic diode test occurs in which
30     !the test temperatures are entered automatically
40     !
50 Viewports:Crt=1          ! Key definitions
60     Printer=701
70     Plotter=705
80     !
90 Datapaths:     !Note: Current Source = 708
100               !      Temperature Sensor = 706
110               !      @Current Meter = 712 or 714
120               !      @Voltage Meter = 713 or 715
130    !I/O pathname assigned to keyboard.
140    ASSIGN @Keyboard TO 2
150    !Non ASCII code to clear CRT
160    Clear_screen$=CHR$(255)&CHR$(75)
170    !I/O pathname assigned to digital thermometer
180    ASSIGN @Thermometer TO 706
190 Information: !
200    PRINTER IS Crt         !Should be Crt
210    PRINT
220    PRINT
230    PRINT
240    PRINT
250    !
260    !Print TAB(X) causes printing to begin at column X of the CRT or paper
270    !
280    PRINT TAB(34);"**  AUTO TEST    **"
290    PRINT
300    PRINT TAB(23);"The folowing program, AUTO TEST, is used"
310    PRINT TAB(18);"to calculate the series resistance, ideality factor,"
320    PRINT TAB(18);"saturation current, and nKT/q ratio, after having"
330    PRINT TAB(18);"measured the voltages corresponding to specified"
340    PRINT TAB(18);"currents in the range 1 microamp - 3 milliamps."
350    PRINT TAB(18);"AUTO TEST repeats this cycle everytime a new initial"
360    PRINT TAB(18);"temperature is entered(ie. T1 changes)."
370    PRINT TAB(18);"Once all the measurements have been made, a plot"
380    PRINT TAB(18);"can be made to show the variation of the nKT/q ratio"
390    PRINT TAB(18);"with temperature."
400    PRINT
410    PRINT
420    PRINT TAB(18);"Press CONTINUE to proceed"
430    PAUSE
```

```
440    OUTPUT @Keyboard;Clear_screen$;
450    !
460    !Print TABXY(X<Y) causes printing to begin at column X on line Y
470    !(SEE pg.287 MANUAL #1).
480    !CRT=18 lines  PAPER=66 lines at 6 lines per inch.
490    !
500    PRINT TABXY(35,5);"EQUIPMENT"
510    PRINT TABXY(24,7);"CURRENT SOURCE: KEITHLEY 220 "
520    PRINT TABXY(24,8);"MEASURING DEVICE: HP3456A DIGITAL VOLTMETER OR"
530    PRINT TABXY(42,9);"HP3478A MULTIMETER"
540    PRINT TABXY(24,10);"TEMPERATURE SENSOR: LAKESHORE DRC-80 THERMOMETER"
550    PRINT TABXY(24,11);"PRINTER: HP82905B External"
560    PRINT TABXY(24,12);"PLOTTER: HP7470A External"
570    PRINT TABXY(24,16);"Press CONTINUE to proceed"
580    PAUSE
590 Begin:  !
600    OUTPUT @Keyboard;Clear_screen$;      !Clear the screen
610    DISP "ENTER YOUR INITIALS";
620    INPUT Initials$                      !Initials are typed in
630    DISP "ENTER TODAY'S DATE(DD,MMMM,YYYY)";
640    INPUT D2,M$,Y4
650    DISP "ENTER DIODE NUMBER";
660    !Diode number is typed in (no more than 18 characters)
670    INPUT D$
680    DISP "CHOOSE SCALE        1=Regular     2=Transistor    3=New          ";
690    INPUT H                              !Number of choice is enterd
700    IF H=1 THEN H$="Reg  0.6-1.2 V"      !Range is 0.6 to 1.2 volts
710    IF H=2 THEN H$="Trans  0.3-0.9 V"    !Range is 0.3 to 0.9 volts
720    IF H=3 THEN
730      DISP "ENTER MINIMUM VOLTAGE";
740      INPUT H1                           !Range is input voltage to input
                                             voltage plus 0.6.
750      X$="New   Vmin="
760      Y$=VAL$(H1)
770      H=H1
780      H$=X$&Y$
790    END IF
800    DISP "CHOOSE MEASURING DEVICE   1=3456A Voltmeter  2=3478A Multimeter  ";
810    INPUT M                              !Number of choice is entered.
820    IF M=1 THEN
830      K$="HF1R1FL1T4"                    !See explanation of instrument code.
840    !Assignment of I/O pathnames.(see PG.251 manual #1)
850      ASSIGN @Meters TO 712,713
860      ASSIGN @Current_meter TO 712
870      ASSIGN @Voltage_meter TO 713
880      K1$="HP3456A - accuracy"
890    END IF
```

```
910      K$="HON5"                               !See explanation of instrument code.
920    !Assignment of I/O pathnames.(see PG.251 manual #1)
930      ASSIGN @Meters TO 714,715
940      ASSIGN @Current_meter TO 714
950      ASSIGN @Voltage_meter TO 715
960      K1$="HP3478A - speed"
970    END IF
980    IF M>2 OR M<1 THEN                        !-------------------------------
990      DISP "IMPROPER ENTRY,TRY AGAIN"         !Prevents acceptance of entries
1000     WAIT 1                                  !other than 1 or 2
1010     GOTO 800                                !
1020   END IF                                    !-------------------------------
1030 Double_check:PRINTER IS Crt
1040   OUTPUT @Keyboard;Clear_screen$;      !Clears the screen
1050   PRINT TABXY(25,9);"** Double Check Information **"
1060   PRINT TABXY(29,11);"TEST DATE: ";D2;" ";M$;" ";Y4
1070   PRINT TABXY(29,12);"DIODE NUMBER: ";D$
1080   PRINT TABXY(29,13);"SCALE: ";H$
1090   PRINT TABXY(29,14);"MEASURING DEVICE: ";K1$
1100   DISP "ANY CHANGES      (YES OR NO)";
1110   INPUT Changes$                            !YES or NO is typed in
1120   IF Changes$="YES" OR Changes$="yes" THEN
1130     PRINT "What Do You Wish To Change?"
1140     PRINT "ENTER X:   0-NOTHING  1-DATE  2-DIODE NUMBER  3-SCALE  4-DEVICE
5-ALL"
1150     INPUT Change                            !Number of choice is entered.
1160     GOTO Changes
1170   ELSE
1180     GOTO Equipment_check
1190   END IF
1200   !Case for 0 change
1210 Changes:IF Change=0 THEN GOTO Equipment_check
1220   !Case for changing just the date.
1230 Datechange:IF Change=1 THEN
1240     DISP "ENTER NEW DATE  (DD,MMMM,YYYY)";
1250     INPUT D2,M$,Y4
1260   END IF
1270   !Case for changing just the diode number
1280 Diodechange:IF Change=2 THEN
1290     DISP "ENTER NEW DIODE NUMBER";
1300     INPUT D$
1310   END IF
```

```
1320  !Case for changing just the plotting scale.
1330 Scalechange:IF Change=3 THEN
1340    DISP "ENTER NEW SCALE  1=Regular   2=Transistor  3=New";
1350    INPUT H
1360    IF H=1 THEN H$="Reg  0.6 - 1.2 V"
1370    IF H=2 THEN H$="Trans  0.3 - 0.9 V"
1380    IF H=3 THEN
1390      DISP "ENTER NEW MINIMUM VOLTAGE ";
1400      INPUT H1
1410      X$="New    Vmin="
1420      Y$=VAL$(H1)
1430      H=H1
1440      H$=X$&Y$
1450    END IF
1460  END IF
1470  !Case for changing just the measuring device.
1480 Devicechange:IF Change=4 THEN
1490    DISP "CHOOSE NEW DEVICE   1=3456A Voltmeter  2=3478A Multimeter  ";
1500    INPUT M
1510    IF M=1 THEN
1520      K$="HF1R1FL1T4"
1530      ASSIGN @Meters TO 712,713
1540      ASSIGN @Current_meter TO 712
1550      ASSIGN @Voltage_meter TO 713
1560      K1$="HP3456A - accuracy"
1570    END IF
1580    IF M=2 THEN
1590      K$="HON5"
1600      ASSIGN @Meters TO 714,715
1610      ASSIGN @Current_meter TO 714
1620      ASSIGN @Voltage_meter TO 715
1630      K1$="HP3478A - speed"
1640    END IF
1650  END IF
1660  !Case for changing everything.
1670 Change_all:IF Change=5 THEN GOTO Begin
1680  GOTO Double_check
1690 Equipment_check:OUTPUT @Keyboard;Clear_screen$;
1700  PRINT TABXY(16,9);"ARE YOU SURE ALL INSTRUMENTS ARE ON?   (YES OR NO)"
1710  INPUT Answer$                        !YES or NO is typed in.
1720  OUTPUT @Keyboard;Clear_screen$;
1730  IF Answer$="NO" OR Answer$="no" THEN
1740    PRINT TABXY(16,9);"CHECKING - the printer will ring a buzzer,"
1750    PRINT TABXY(27,10);"a current, voltage, and temperature reading"
1760    PRINT TABXY(27,11);"should appear, and the plotter will pick up a"
1770    PRINT TABXY(27,12);"pen and replace it."
1780    PRINT TABXY(21,14);"A message - ALL INSTRUMENTS ARE ON - will appear."
1790    PRINT TABXY(16,17);"Press CONTINUE to proceed"
1800    PAUSE
```

```
1810     OUTPUT @Keyboard;Clear_screen$;
1820     REMOTE 7                        !All instruments controlled on HPIB.
1830     OUTPUT @Meters;K$               !Instrument code sent to measuring
                                          device.
1840     OUTPUT 708;"ROP1W1V4F1X"        !See explanation of instrument code.
1850     PRINTER IS Printer
1860     PRINT CHR$(7)          !Ring Printer's buzzer
1870     PRINTER IS Crt
1880     OUTPUT 708;"I1E-6X"             !Current source puts out 10^-6 A.
1890     TRIGGER @Meters                 !Measuring devices are triggered.
1900     ENTER @Current_meter;Current !Device 712 or 714 reads current.
1910     ENTER @Voltage_meter;Voltage !Device 713 or 715 reads voltage.
1920     ENTER @Thermometer;P$,T$        !Device 706 temperature.
1930     Temperature=VAL(T$[2])
1940    !
1950    !Temperature=VAL(T$[2]) : The temperature read by the digital
1960    !thermometer is an alphanumeric variable(string) which looks
1970    !like 00290.0 . The VAL(T$[X]) function converts the string
1980    !into a numeric value,beginning at the XTH location of the
1990    !string.  (see chapter 5,manual #1)
2000    !
2010     PRINT TABXY(10,16);"Current= ";Current,"Voltage= ";Voltage,"Temperature=
 ";Temperature
2020     PRINTER IS Plotter
2030     PRINT "IN;PU;SP1;SP0"        !Pick up pen, put pen down
2040     PRINTER IS Crt
2050     PRINT TABXY(29,17);"ALL INSTRUMENTS ARE ON!"
2060     WAIT 5
2070     GOTO Equipment_check
2080   ELSE
2090     GOTO Print_info
2100   END IF
2110 Print_info:PRINTER IS Printer    !Should be Prinetr for hard copy of results
2120   PRINT TAB(14);CHR$(27);"&k1S** AUTOMATIC DIODE TEST **";CHR$(27);"&k0S"
2130   FOR W=1 TO 3                    !------------------------------
2140     PRINT                         !Skips three lines on printer.
2150   NEXT W                          !------------------------------
2160   PRINT USING "8X,11A,2D,1X,4A,1X,4D";"TEST DATE: ";D2;M$;Y4
2170   PRINT TAB(9);"DIODE NUMBER: ";D$
2180   PRINT TAB(9);"MEASURING DEVICE: ";K1$
2190   PRINT TAB(9);"OPERATOR: ";Initials$
2200   PRINT
2210   OPTION BASE 0                !Sets lower bound of array subscripts to zero.
2220   REMOTE 7                     !All instruments controlled on HPIB.
2230   OUTPUT @Meters;K$            !Instrument code sent to measuring devices.
2240   OUTPUT 708;"ROP1W1V4F1X"     !See explanation of instrument code.
2250 Choices:                       !Menu keys are defined.
```

```
2260  PRINTER IS Crt       !Should be Crt
2270  PRINT TAB(8);"*See Keys at upper left corner of Keyboard"
2280  PRINT
2290  PRINT TAB(8);"Key  Label          Result"
2300  PRINT TAB(8);"k0   SINGLE CYCLE   1. Point Resistance is given"
2310  PRINT TAB(28);"2. One measurement cycle for one temperature"
2320  PRINT TAB(28);"3. One full LOG I vs. V plot"
2330  PRINT
2340  PRINT TAB(8);"k1   COMPLETE TEST  1. Point Resistance is measured"
2350  PRINT TAB(28);"2. A measurement cycle occurs for specific"
2360  PRINT TAB(31);"temperatures in the range 300k-20k"
2370  PRINT TAB(28);"3. 4 full LOG I vs. V plots and a VO vs. TEMP plot"
2380  PRINT
2390  PRINT TAB(8);"k2   SKIP SHORT     1. Point Resistance is given"
2400  PRINT TAB(28);"2. A measurement cycle occurs for specific"
2410  PRINT TAB(31);"temperatures in the range 300k-20k"
2420  PRINT TAB(28);"3. 4 full LOG I vs. V plots and a VO vs. TEMP plot"
2430  PRINT
2440  PRINT TAB(8);"k3   STOP TESTING   End of Testing"
2450  DISP "CHOOSE TEST DESIRED:press key"
2460 !Keys
2470  ON KEY 0 LABEL "SINGLE CYCLE" GOTO Single     !------------------
2480  ON KEY 1 LABEL "COMPLETE TEST" GOTO Complete !
2490  ON KEY 2 LABEL "SKIP SHORT" GOTO Skipshort   !Produces softkey
2500  ON KEY 3 LABEL "STOP TESTING" GOTO End       !menu and maintains
2510  ON KEY 4 LABEL "" GOTO 2450                  !display until a
2520  ON KEY 5 LABEL "" GOTO 2450                  !selection has been
2530  ON KEY 6 LABEL "" GOTO 2450                  !made.
2540  ON KEY 7 LABEL "" GOTO 2450                  !
2550  ON KEY 8 LABEL "" GOTO 2450                  !
2560  ON KEY 9 LABEL "" GOTO 2450                  !
2570 Spin:GOTO Spin                                !------------------
2580  LOCAL LOCKOUT 7
2590 Complete:  !COMPLETE TEST begins here
2600  OFF KEY                            !Causes menu to disappear
2610  OUTPUT @Keyboard;Clear_screen$;    !Clears screen of definitions.
2620  !P=1 signals a series of cycles dependent on temperature
2630  !will follow.
2640  !P=0 signals that only one set of measuments will follow
2650  !ie.cycle 0.
2660  !
2670  P=1                                !Sets number of cycles.
2680 Short:DISP "SHORT DIODE LEADS- a beep signals a proper short,";
2690  DISP "when shorted, press CONTINUE"
2700  PAUSE
```

```
2710  OUTPUT 708;"I2E-3X"          !Current source puts out 2*10^-3 A.
2720  TRIGGER @Meters
2730  ENTER @Current_meter;J9
2740  ENTER @Voltage_meter;V9
2750  IF V9>.05 THEN
2760    DISP "IMPROPER SHORT - try again",
2770    WAIT 2
2780    GOTO Short
2790  ELSE
2800    BEEP 1000,.5
2810  END IF
2820  !
2830  !The current meter (device #712 or #714) reads voltage
2840  !that are converted to current values by OHM'S LAW i.e.
2850  !I=V/R  or  I=(V/100)*1000=V*10
2860  !
2870  J9=ABS(J9*10)             !Voltage reading changed to current.
2880  R9=V9/J9*1000             !Point resistance is calculated.
2890  PRINTER IS Printer        !Should be Printer for hard copy of results
2900  PRINT USING "8X,19A,D.2D,5A";"POINT RESISTANCE= ";R9;" OHMS"
2910  GOTO 3040        !Goes to lowering the current to zero
2920 Single: !SINGLE CYCLE begins here
2930  OFF KEY                            !Causes menu to disappear.
2940  OUTPUT @Keyboard;Clear_screen$;    !Clears screen of definitions.
2950  P=0                                !Sets number of cycles.
2960  GOTO 3010      !Goes to "ENTER WHISKER RESISTANCE"
2970 Skipshort:  !Test with NO SHORT begins here
2980  OFF KEY                            !Causes menu to disappear.
2990  OUTPUT @Keyboard;Clear_screen$;    !Clears screen of definitions.
3000  P=1                                !Sets number of cycles.
3010  DISP "ENTER WHISKER RESISTANCE";
3020  INPUT R9         !Value of whisker resistance is typed in.
3030  GOTO 2890        !Goes to printing the value of the resistance
3040  OUTPUT 708;"I0E-3X"   !Lowers the current to zero
3050  WAIT .3
3060  TRIGGER @Current_meter
3070  ENTER @Current_meter;J
3080 Contact:DISP "CONNECT DIODE - a beep indicates proper contact, ";
3090  DISP "when connected press CONTINUE"
3100  PAUSE
```

```
3110  OUTPUT 708;"I25E-6X"        !Current source produces 2.5*10^-6 A.
3120  TRIGGER @Voltage_meter
3130  ENTER @Voltage_meter;C1
3140  WAIT .1
3150 !If C1 is between .2 and 3.6 volts then proper contact has been made
3160  L1=C1>.2 AND C1<3.6
3170  IF L1<>1 THEN
3180    DISP "IMPROPER CONTACT - try again",
3190    WAIT 2
3200    GOTO Contact
3210  ELSE
3220    BEEP 1000,.5
3230  END IF
3240  DISP "                                        "
3250  PRINTER IS Crt          !Should be Crt
3260  PRINT TABXY(35,5);"THE TEST IS IN PROGRESS"
3270  PRINT TABXY(29,7);"The testing shall proceed automatically"
3280  PRINT TABXY(24,8);"for the next 2.5 hours (time approximate) at"
3290  PRINT TABXY(24,9);"which point it is possible to plot VO vs. TEMP."
3300  PRINT TABXY(29,11);"If a SINGLE CYCLE has been selected"
3310  PRINT TABXY(24,12);"the test will end in approximately 3 minutes."
3320  PRINT TABXY(24,15);"If a mistake has been made press RESET then RUN"
3330 Memory:  ! Memory Allocation
3340  DIM I(15)       !Ideal Currents
3350  DIM C(15)       !Ideal Current in mA
3360  DIM I1(15)
3370  DIM I2(15)
3380  DIM J1(15)      !Measured Currents
3390  DIM J2(15)
3400  DIM J3(15)
3410  DIM J4(15)
3420  DIM V1(15)      !Measured Voltages
3430  DIM V2(15)
3440  DIM V3(15)
3450  DIM V4(15)
3460  DIM V(15)       !Interpolated Voltages
3470  DIM V6(15)
3480  DIM T1(60)      !Measured Temperatures
3490  DIM T2(60)
3500  DIM T3(60)
3510  DIM Temp(15)
3520  DIM VO(60)      !nKT/q
```

```
3530 Currents:  !Ideal Currents are calculated in 10^-6 A.
3540  FOR M=1 TO 15 STEP 4
3550    I(M)=10^(-7+(M+3)/4)
3560  NEXT M
3570  FOR M=2 TO 14 STEP 4
3580    I(M)=2*10^(-7+(M+2)/4)
3590  NEXT M
3600  FOR M=3 TO 15 STEP 4
3610    I(M)=3*10^(-7+(M+1)/4)
3620  NEXT M
3630  FOR M=4 TO 12 STEP 4
3640    I(M)=6*10^(-7+M/4)
3650  NEXT M
3660  Desired_temp=290          !Variable for comparison
                                 (see detailed explanation)
3670 Majorloop: !Major Loop: EVERYTHING IS DONE AUTOMATICALLY
3680  ON ERROR RECOVER Error_routine   !Activates error routine.
3690  IF P=0 THEN GOTO 3770            !Determines number of cycles.
3700  INTEGER P              !Speeds program execution.(see PG.351,manual #1)
3710  !
3720  !In the event of an error,an extra cycle is indroduced,to cover for
3730  !this,60 cycles can be executed if necessary,before the testing will
3740  !end.Otherwise,approximately 28 cycles are executed.
3750  !
3760  FOR P=1 TO 60          !Maximum number is number of cycles acc. to temps
3770    PRINTER IS Printer    !Should be Printer for hard copy of results
3780    PRINT
3790    PRINT TAB(9);"CYCLE# ";P
3800    IF P=0 THEN
3810    ENTER @Thermometer;P$,T$
3820    T1(P)=VAL(T$[2])
3830    DISP "INITIAL TEMP IS";T1(P)
3840    WAIT 1
3850    GOTO 3940
3860    END IF
3870    PRINT
3880 Temperature1:REPEAT      !See detailed explanation.
3890      ENTER @Thermometer;P$,T$
3900      T1(P)=VAL(T$[2])     !Changes the string into a number
3910      DISP "INITIAL TEMP IS ";T1(P)
3920      WAIT 1
3930    UNTIL T1(P)<=291.50   !Setting the initial temperature
3940    PRINT
3950    PRINT TAB(9);"Initial Temp is ";T1(P);" K"
3960    PRINT
3970    PRINT
3980    INTEGER Q             !Speeds program execution.(see PG.351,manual #1)
```

```
3990  !See detailed explanation.
4000 Measurements:    !Loop: Current and Voltage Measurements
4010    FOR Q=1 TO 15
4020      C(Q)=I(Q)*1.E+3         !Ideal Current changed to mA
4030      OUTPUT 708;"I";I(Q);"X"
4040      TRIGGER @Meters
4050      ENTER @Current_meter;J1(Q)
4060      ENTER @Voltage_meter;V1(Q)
4070      ENTER @Thermometer;P$,T$
4080      Temp(Q)=VAL(T$[2])
4090      J1(Q)=ABS(J1(Q)*10)           !Measured Current changed to mA
4100      DISP "CYCLE: #";P,"TEMP= ";Temp(Q),"LOOP #";Q,"J=";J1(Q);" V=";V1(Q)
4110      D1=(J1(Q)-C(Q))*1.E-3
4120      I2(Q)=I(Q)-(D1*2)       !Ideal Current +/- Offset
4130      OUTPUT 708;"I";I2(Q);"X"
4140      TRIGGER @Meters
4150      ENTER @Current_meter;J2(Q)
4160      ENTER @Voltage_meter;V2(Q)
4170      ENTER @Thermometer;P$,T$
4180      Temp(Q)=VAL(T$[2])
4190      J2(Q)=ABS(J2(Q)*10)     !Current in mA
4200      DISP "CYCLE: #";P,"TEMP= ";Temp(Q),"LOOP #";Q,"J=";J2(Q);" V=";V2(Q)
4210          !Interpolation
4220      IF J1(Q)-J2(Q)=0 THEN
4230        D3=(V1(Q)-V2(Q))/2
4240      ELSE
4250        D3=(V1(Q)-V2(Q))/(J1(Q)-J2(Q))
4260      END IF
4270      V(Q)=V2(Q)+D3*(C(Q)-J2(Q))
4280    NEXT Q
4290    WAIT 1
4300  !Mid-test temperature.
4310 Temperature2:ENTER @Thermometer;P$,T$
4320    T2(P)=VAL(T$[2])          !Changes the string into a number
4330  !See detailed explanation.
4340 Hysterisis:  !Loop: Hysterisis Measurements
4350    OUTPUT 708;"ROP1W1V4F1X"
4360    FOR Q=15 TO 1 STEP -1
4370      OUTPUT 708;"I";I(Q);"X"
4380      TRIGGER @Meters
4390      ENTER @Current_meter;J3(Q)
4400      ENTER @Thermometer;P$,T$
4410      Temp(Q)=VAL(T$[2])
4420      ENTER @Voltage_meter;V3(Q)
4430      J3(Q)=ABS(J3(Q)*10)        !Measured Current changed to mA
4440      DISP "CYCLE: #";P,"TEMP= ";Temp(Q),"LOOP #";Q,"J=";J3(Q);" V=";V3(Q)
```

```
4440       DISP "CYCLE: #";P,"TEMP= ";Temp(Q),"LOOP #";Q,"J=";J3(Q);" V=";V3(Q)
4450       D4=(J3(Q)-C(Q))*1.E-3
4460       I2(Q)=I(Q)-(D4*2)          !Current +/- Offset
4470       OUTPUT 708;"I";I2(Q);"X"
4480       TRIGGER @Meters
4490       ENTER @Current_meter;J4(Q)
4500       ENTER @Voltage_meter;V4(Q)
4510       ENTER @Thermometer;P$,T$
4520       Temp(Q)=VAL(T$[2])
4530       J4(Q)=ABS(J4(Q)*10)        !Current changed to mA
4540       DISP "CYCLE: #";P,"TEMP= ";Temp(Q),"LOOP #";Q,"J=";J4(Q);" V=";V4(Q)
4550          !Interpolation
4560       IF J3(Q)-J4(Q)=0 THEN
4570         D5=(V3(Q)-V4(Q))/2
4580       ELSE
4590         D5=(V3(Q)-V4(Q))/(J3(Q)-J4(Q))
4600       END IF
4610       V6(Q)=V4(Q)+D5*(C(Q)-J4(Q))
4620     NEXT Q
4630     DISP "                              "      !Clears display line.
4640     WAIT 1
4650   !Final temperature of cycle.
4660 Temperature3:ENTER @Thermometer;P$,T$
4670     T3(P)=VAL(T$[2])          !Changes the string into a number
4680     PRINT USING "16X,13A,7X,5A,9X,10A";"CURRENT IN mA";"VOLTS";"HYSTERISIS"
4690     PRINT
4700   !Interpolated voltage and ideal currents are printed
        in the specified format.
4710     FOR K=1 TO 15
4720       PRINT USING 4730;I(K);V(K);V6(K)
4730       IMAGE 17X,D.4DE,8X,Z.5D,9X,Z.5D
4740     NEXT K
4750     PRINT
4760     PRINT
4770     PRINT TAB(9);"Temp During Test is  ";T2(P);" K"
4780     PRINT
4790     PRINT
4800 Calculations:  !SERIES RESISTANCE,ETA-n,SATURATION CURRENT, AND VO
4810     K1=(C(13)-C(9))*LOG(C(9)/C(5))-(C(9)-C(5))*LOG(C(13)/C(9))
4820     K2=(C(13)-C(9))*(V(9)-V(5))-(C(9)-C(5))*(V(13)-V(9))
4830     K3=(C(13)-C(9))*LOG(C(9)/C(5))-(C(9)-C(5))*LOG(C(13)/C(9))
4840     K4=(C(13)-C(9))*(V6(9)-V6(5))-(C(9)-C(5))*(V6(13)-V6(9))
4850     K=K1/K2
4860     K5=K3/K4
4870     Z1=1.6*10^(-19)                !Charge on an electron
4880     B1=1.38*10^(-23)               !Boltzman's Constant
4890     T=(T1(P)+T2(P)+T3(P))/3        !Average temperature during the cycle
4900     E1=Z1/(K*B1*T)                 !ETA - n
4910     E2=Z1/(K5*B1*T)                !Hysterisis ETA - (H)n
```

```
4920     R1=(V(13)-V(9))*LOG(C(9)/C(5))-(V(9)-V(5))*LOG(C(13)/C(9))
4930     R2=(C(13)-C(9))*LOG(C(9)/C(5))-(C(9)-C(5))*LOG(C(13)/C(9))
4940     R=ABS(R1/R2*1000-R9)              !Series Resistance
4950     L6=LOG(C(9))-K*(V(9)-R/1000*C(9))
4960     IO=EXP(L6)                        !Saturation Current
4970     VO(P)=E1*B1*T/Z1*1000             ! nKT/q
4980   !Calculated values are printed.
4990     PRINT USING "16X,7A,2Z.2D,5A";"R     =";R;" OHMS"
5000     PRINT USING "16X,7A,3D.3D";"n     =";E1
5010     PRINT USING "16X,7A,3D.3D";"H(n) =";E2
5020     PRINT USING "16X,7A,D.3DE,4A";"Io   =";IO;" mA"
5030     PRINT USING "16X,7A,3D.2D,3A";"Vo   =";VO(P);" mV"
5040     PRINT
5050     PRINT
5060     PRINT TAB(9);"Final Temp is ";T3(P);" K"
5070 Check_for_plot:     !
5080     IF P=0 THEN       !Single cycle gets one plot.
5090   !The message is printed and the paper advance to the next sheet.
5100       PRINT TAB(8);"*See PLOT";CHR$(12)
5110       GOTO Plot1      !LOG I vs V plot.
5120     END IF
5130     IF P=1 THEN            !Room Temperature Plot
5140   !CHR$(12) is a non ASCII code for pressing the formfeed button
        on the printer,internally.
5150       PRINT TAB(8);"*See PLOT";CHR$(12)
5160       GOTO Plot1
5170     END IF
5180     IF T2(P)>=195 AND T2(P)<=205 THEN      !Desired temp. is 200K.
5190       PRINT TAB(8);"*See PLOT";CHR$(12)
5200       GOTO Superplots                      !Superimposed plot.
5210     END IF
5220     IF T2(P)>=95 AND T2(P)<=105 THEN       !Desired temp. is 100K.
5230       PRINT TAB(8);"*See PLOT";CHR$(12)
5240       GOTO Superplots
5250     END IF
5260     IF T2(P)<=20 THEN                      !Desired temp. is 20.
5270       PRINT TAB(8);"*See PLOT";CHR$(12)
5280       GOTO Superplots
5290     END IF
5300     IF T2(P)<=20 THEN          !Checks for the last temperature.
5310       PRINT CHR$(12)           !Advance paper (formfeed).
5320       GOTO Endloop             !Regardless of the value of P,if
                                     T2(P)=20K then the measuments stop.
5330     END IF
5340     PRINT CHR$(12)             !advance paper (formfeed).
```

```
5350  !See detailed explanation.
5360 Temperatureloop:  !Determining the next initial temperature
5370    Delta_temp=T2(P)-Desired_temp
5380    IF ABS(Delta_temp)<=.2 THEN
5390      REPEAT
5400        ENTER @Thermometer;P$,T$
5410        TO=VAL(T$[2])  !Changes the string into a number
5420        DISP "WAITING FOR NEXT TEMPERATURE ";TO
5430        WAIT 1
5440      UNTIL TO<=T1(P)-10
5450    ELSE
5460      REPEAT
5470        ENTER @Thermometer;P$,T$
5480        TO=VAL(T$[2])  !Changes the string into a number
5490        DISP "WAITING FOR NEXT TEMPERATURE ";TO
5500        WAIT 1
5510      UNTIL TO<=T1(P)-10+ABS(Delta_temp)
5520    END IF
5530    Desired_temp=Desired_temp-10
5540  NEXT P                          !Triggers beginning of new cycle.
5550 Endloop:  !End of Major Loop
5560  OFF ERROR                       !Deactivates error routine.
5570  PRINTER IS Printer
5580  PRINT USING "8X,11A,2D,1X,4A,1X,4D";"TEST DATE: ";D2;M$;Y4
5590  PRINT TAB(9);"DIODE NUMBER: ";D$
5600  PRINT
5610  PRINT
5620  PRINT
5630  PRINT USING "29X,9A,5X,8A";"TEMP IN K";"VO IN mV"
5640  PRINT
5650 !Mid-test temperature and VO values are printed in the
      specified format.
5660  FOR K=1 TO 30
5670    PRINT USING 5680;T2(K);VO(K)
5680    IMAGE 31X,3D.D,8X,M2D.2D
5690  NEXT K
5700  PRINT CHR$(12)
5710  PRINT USING "8X,11A,2D,1X,4A,1X,4D";"TEST DATE: ";D2;M$;Y4
5720  PRINT TAB(9);"DIODE NUMBER: ";D$
5730  PRINT TAB(9);"PAGE 2"
5740  PRINT
5750  PRINT
5760  PRINT
5770  PRINT USING "29X,9A,5X,8A";"TEMP IN K";"VO IN mV"
5780  PRINT
```

```
5790 !Mid-test temperatures and VO values are printed in the
      specified format.
5800  FOR K=31 TO 60
5810    PRINT USING 5820;T2(K);VO(K)
5820    IMAGE 31X,3D.D,8X,M2D.2D
5830  NEXT K
5840  GOTO Plot2                       !VO vs T2(P) plot.
5850 Plot1:  !Log I vs. V Routine
5860  IF H=1 THEN G=.5            !Scale Parameters
5870  IF H=2 THEN G=.2
5880  IF H=H1 THEN G=H-.1
5890  PRINTER IS Plotter
5900  PRINT "IN;SP1;IP1200,850,9250,7600"
5910  PRINT "SC1,7,0,4"
5920  PRINT "PU1,0,PD7,0,7,3.5,1,3.5,1,0,PU"
5930  PRINT "SI.2,.3;TL87.5"
5940  FOR M=1 TO 7
5950    PRINT "PA";M,",0;XT,"
5960    A$=VAL$(M/10+G)
5970    PRINT "CP-1.0,-1;LB";A$;""
5980  NEXT M
5990  PRINT "PA4.0,0;CP-3,-2.5;LBVOLTS"
6000  FOR N=0 TO 3
6010    PRINT "TL100"
6020    PRINT "PA1,",N,"YT;"
6030    B$=VAL$(10^(N-3))
6040    PRINT "CP-4,-.25;LB";B$;""
6050  NEXT N
6060  FOR K=0 TO 3
6070    PRINT "PA1,",K+.47712,"YT;"
6080    C$=VAL$(3*10^(K-3))
6090    PRINT "CP-4,-.25;LB";C$;""
6100  NEXT K
6110  PRINT "PA1,3.5 SI.25,.35 CP0,2.0"
6120  PRINT "LBDEVICE: ";D$;""
6130  PRINT "CP22,0"
6140  PRINT "LBDATE: ";M$;D2;",";Y4;""
6150  PRINT "PA1,1.75 SI.2,.3 CP-6,-3"
6160  PRINT "DI0,1;LBLOG I in mA DI2,0;PA1,0"
6170 Superplots:               !Superimposed Plots
6180  IF H=1 THEN G=.5        !Scale Parameters
6190  IF H=2 THEN G=.2
6200  IF H=H1 THEN G=H-.1
```

```
6210  PRINTER IS Plotter
6220  PRINT "SC1,7,0,4"
6230  PRINT "SP1"
6240  DIM X(15)
6250  DIM Y(15)
6260  X(1)=10*(V(1)-G)
6270  Y(1)=LGT(C(1))+3
6280  PRINT "PU";X(1);Y(1)
6290  FOR M=2 TO 15
6300    X(M)=10*(V(M)-G)
6310    Y(M)=LGT(C(M))+3
6320    PRINT "PD";X(M);Y(M)
6330  NEXT M
6340  PRINT "SP2"
6350  FOR M=15 TO 1 STEP -1
6360    X(M)=10*(V6(M)-G)
6370    Y(M)=LGT(C(M))+3
6380    PRINT "PD";X(M);Y(M)
6390  NEXT M
6400  PRINT "PU;SP0"
6410  PRINT
6420  PRINT
6430  IF P=0 THEN GOTO Begin
6440  GOTO 5300
6450 Plot2:DISP "CHANGE PLOTTER PAPER - THEN PRESS CONTINUE"
6460  PAUSE
6470  !V0 vs Temperature Plot
6480  PRINTER IS Plotter
6490  PRINT "IN;SP1;IP1500,1200,8950,7650"
6500  PRINT "SC1,7,0,5"
6510  PRINT "PU1,0,PD7,0,7,4,1,4,1,0,PU"
6520  PRINT "SI.2,.3;TL80"
6530  PRINT "PA";1,",0;XT,"
6540  E$=VAL$(0)
6550  PRINT "CP-1.0,-1;LB";E$;""
6560  FOR M=2 TO 7
6570    PRINT "PA";M,",0;XT,"
6580    E$=VAL$((M+M-2)*25)
6590    PRINT "CP-1.0,-1;LB";E$;""
6600  NEXT M
6610  PRINT "PA3.5,0;CP-3,-3.5;LBTEMPERATURE in K"
6620  FOR N=0 TO 4
6630    PRINT "TL100"
6640    PRINT "PA1,",N,"YT;"
6650    F$=VAL$((N+3)*5)
6660    PRINT "CP-4,-.25;LB";F$;""
6670  NEXT N
```

```
6680  PRINT "PA1,4 SI.25,.35 CP0,3.0"
6690  PRINT "LBDEVICE: ";D$;""
6700  PRINT "CP22,0"
6710  PRINT "LBDATE: ";M$;D2;",";Y4;""
6720  PRINT "PA1,2 SI.2,.3 CP-8,-2"
6730  PRINT "DI0,1;LBV0 in mV DI2,0;PA1,.5"
6740  PRINT "SP2"
6750  DIM X1(60)
6760  DIM Y1(60)
6770  X1(1)=(T2(1)/50)+1
6780  Y1(1)=((V0(1)-15)/5)+.5
6790  PRINT "PU";X1(1);Y1(1)
6800  FOR M=2 TO 60
6810    IF T2(M)=0 OR V0(M)=0 THEN GOTO 6860
6820    X1(M)=(T2(M)/50)+1
6830    Y1(M)=((V0(M)-15)/5)+.5
6840    PRINT "PD";X1(M);Y1(M)
6850  NEXT M
6860  PRINT "PU;SP0"
6870  PRINT
6880  PRINT
6890  DISP "CHANGE PLOTTER PAPER - THEN PRESS CONTINUE"
6900  PAUSE
6910  PRINTER IS Printer       !Returning to the external printer
6920  GOTO Choices
6930 Error_routine:E=ERRN
6940  IF E>0 THEN
6950    PRINT "ERROR";ERRN;CHR$(12)
6960    IF P>60 THEN GOTO Endloop
6970    GOTO 5540                        !NEXT P
6980  END IF
6990 End:OUTPUT @Keyboard;Clear_screen$;
7000     DISP "END OF TESTING"
7010     END
```

Appendix D
Flowchart

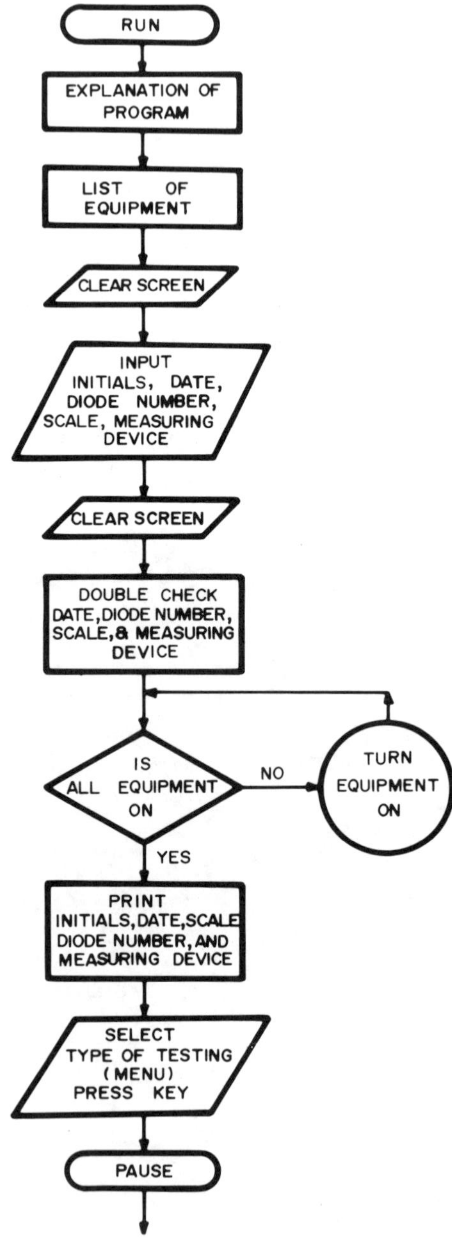

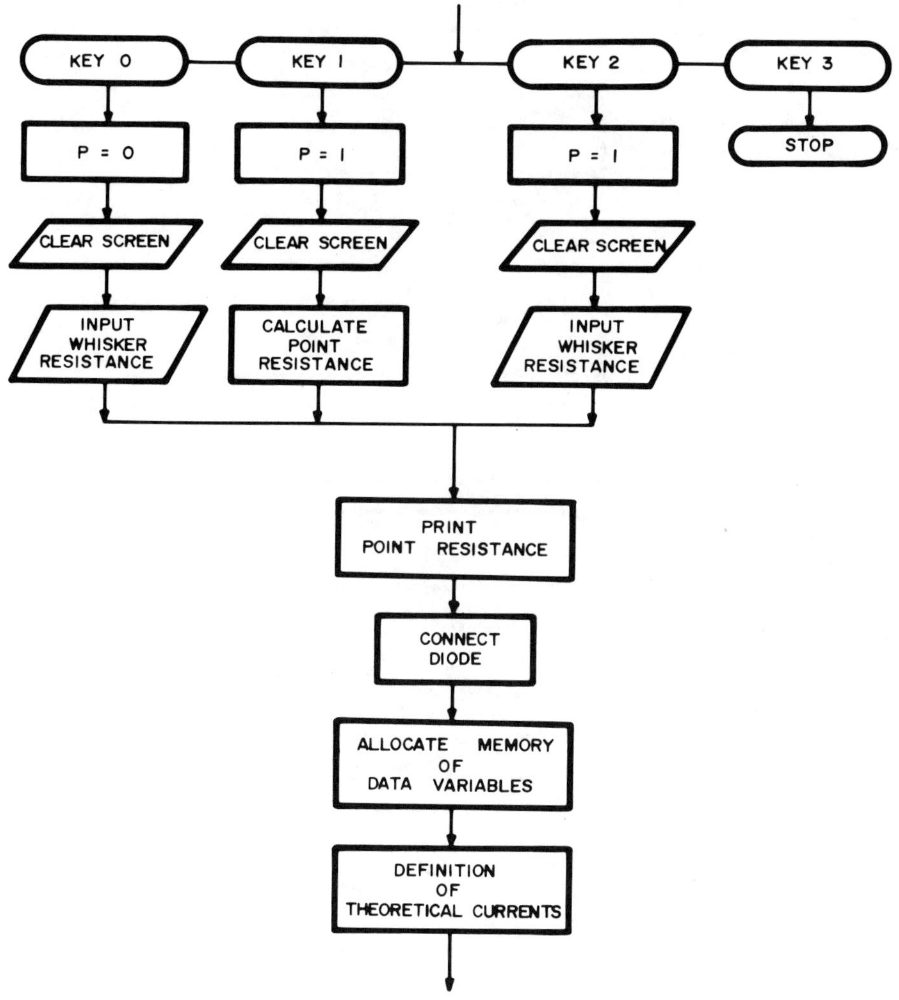
KEY 0
KEY 1
KEY 2
KEY 3
P = 0
P = 1
P = 1
STOP
CLEAR SCREEN
CLEAR SCREEN
CLEAR SCREEN
INPUT WHISKER RESISTANCE
CALCULATE POINT RESISTANCE
INPUT WHISKER RESISTANCE
PRINT POINT RESISTANCE
CONNECT DIODE
ALLOCATE MEMORY OF DATA VARIABLES
DEFINITION OF THEORETICAL CURRENTS

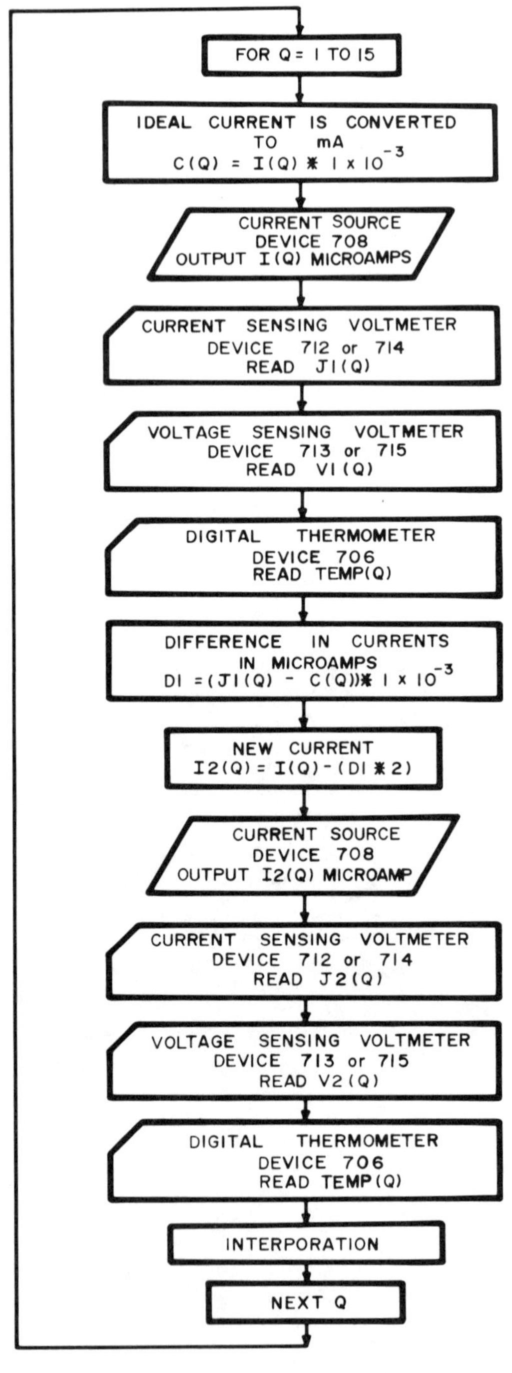
FOR Q = 1 TO 15
IDEAL CURRENT IS CONVERTED TO mA
C(Q) = I(Q) ✱ 1 x 10^-3
CURRENT SOURCE
DEVICE 708
OUTPUT I(Q) MICROAMPS
CURRENT SENSING VOLTMETER
DEVICE 712 or 714
READ J1(Q)
VOLTAGE SENSING VOLTMETER
DEVICE 713 or 715
READ V1(Q)
DIGITAL THERMOMETER
DEVICE 706
READ TEMP(Q)
DIFFERENCE IN CURRENTS IN MICROAMPS
DI = (J1(Q) - C(Q))✱ 1 x 10^-3
NEW CURRENT
I2(Q) = I(Q) - (DI ✱ 2)
CURRENT SOURCE
DEVICE 708
OUTPUT I2(Q) MICROAMP
CURRENT SENSING VOLTMETER
DEVICE 712 or 714
READ J2(Q)
VOLTAGE SENSING VOLTMETER
DEVICE 713 or 715
READ V2(Q)
DIGITAL THERMOMETER
DEVICE 706
READ TEMP(Q)
INTERPORATION
NEXT Q

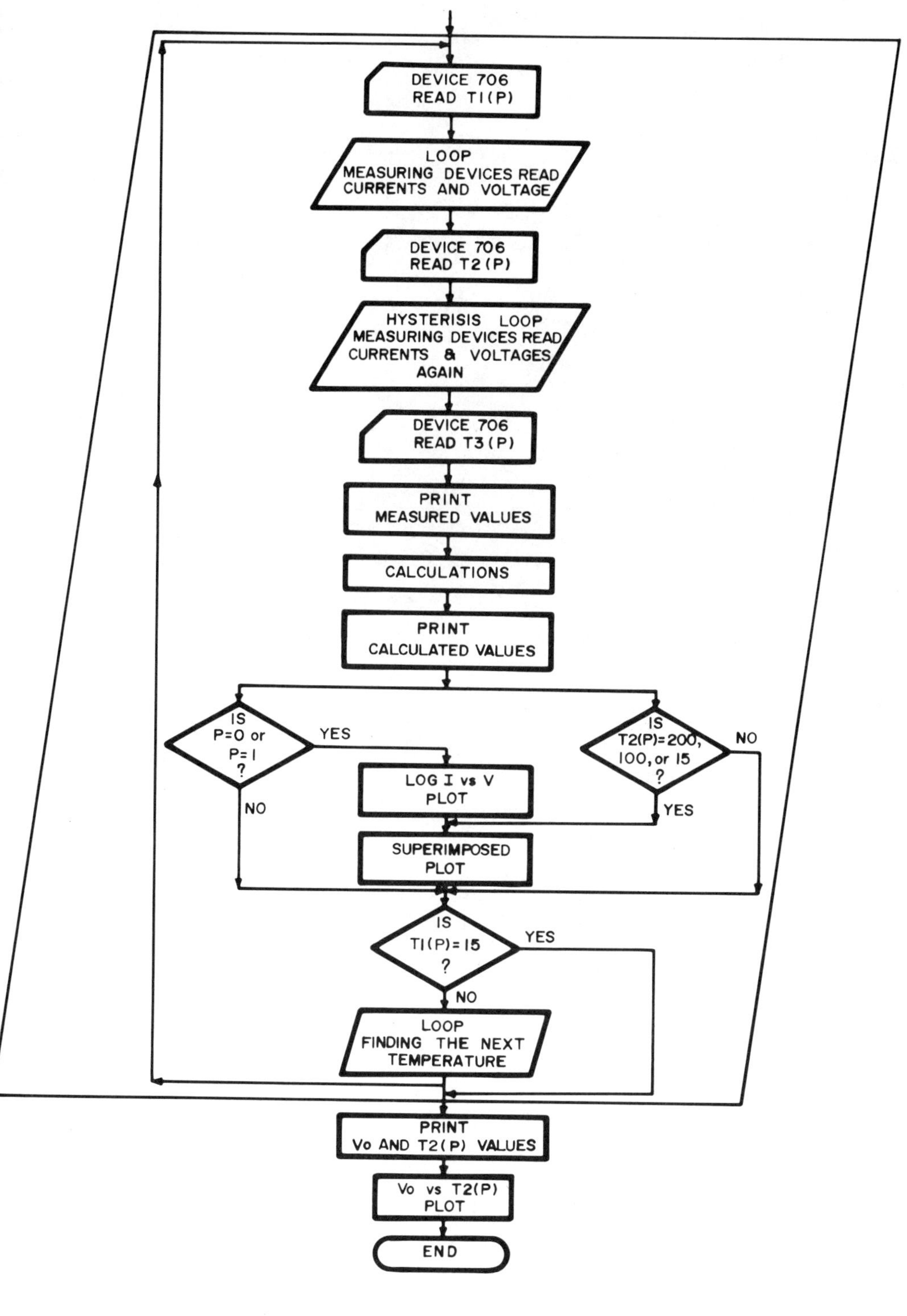
DEVICE 706
READ T1(P)
LOOP
MEASURING DEVICES READ
CURRENTS AND VOLTAGE
DEVICE 706
READ T2(P)
HYSTERISIS LOOP
MEASURING DEVICES READ
CURRENTS & VOLTAGES
AGAIN
DEVICE 706
READ T3(P)
PRINT
MEASURED VALUES
CALCULATIONS
PRINT
CALCULATED VALUES
IS
P=0 or
P=1
?
YES
NO
LOG I vs V
PLOT
IS
T2(P)=200,
100, or 15
?
NO
YES
SUPERIMPOSED
PLOT
IS
T1(P)=15
?
YES
NO
LOOP
FINDING THE NEXT
TEMPERATURE
PRINT
Vo AND T2(P) VALUES
Vo vs T2(P)
PLOT
END

CHAPTER 3

Multichannel Far-Infrared Collective Scattering System for Plasma Wave Studies

D. L. Brower, H. K. Park, W. A. Peebles, N. C. Luhmann, Jr.

Institute of Plasma and Fusion Research
University of California, Los Angeles
Los Angeles, CA 90024

I. Introduction

Thomson scattering of electromagnetic radiation has been established as a valuable diagnostic tool possessing a wide range of applications in both laboratory and controlled thermonuclear fusion plasma research. This nonperturbing diagnostic permits internal measurement of electron temperature and density, ion temperature, magnetic field direction, and electron density fluctuation spectra (Sheffield, 1975). The application to be explored in this chapter concerns the use of millimeter and submillimeter wave collective Thomson scattering (Luhmann and Peebles, 1984, 1985; Luhmann, 1979) to determine the spatial and temporal distribution

ISBN 0-12-147700-2

of electron density fluctuations in a magnetically confined fusion plasma, which is of interest for several reasons. First, low-frequency instabilities driven by plasma inhomogeneities may be responsible for the anomalous transport observed in present day toroidal confinement devices (Waltz et al., 1980; Horton and Estes, 1979). Supplementary heating schemes for thermonuclear plasmas are likely to enhance this microturbulence level, making measurement of the frequency and wavenumber spectra of these fluctuations essential. Second, rf heating at the lower hybrid and ion cyclotron frequencies, in addition to electron cyclotron heating, are anticipated to develop into cost-effective alternatives to neutral beam heating. A study of the launching, propagation, mode conversion, and damping of such waves is of importance in understanding the heating mechanisms and improving efficiencies. Third, the external launching of waves in fusion plasmas and a measurement of their dispersion relations can lead to temporally and spatially resolved determination of plasma parameters such as the ion temperature. Finally, laboratory plasma investigations of nonlinear processes such as stimulated Brillouin scattering (Huey et al., 1980 and 1984; Mase et al., 1981; Clayton et al., 1982), optical mixing (Pawley et al., 1982; Kroll et al., 1964; Cohen et al., 1972 and 1975; Schmidt, 1973; James and Thompson, 1967; Rosenbluth and Liu, 1972; Cohen, 1975 and 1984; Weibel, 1976a,b, 1976b), spontaneous magnetic field generation (Obenschain and Luhmann, 1977 and 1979; DiVergilio et al., 1977; Raven and Rumsby, 1977; Bezzerides et al., 1974), and Langmuir collapse and soliton formation (Kim et al., 1974; Wong, 1976; A. Y. Lee et al., 1982; Morales and Lee, 1977; Chen and Liu, 1977; Rhodes et al., 1979; Elsasser and Schamel, 1977; Schamel and Elsasser, 1978) require a nonperturbing measurement of wave phenomena possessing high spatial resolution. Until recently, measurements such as those mentioned above were performed using Langmuir probes. However, probes can only survive in the low-temperature edge region (limiter shadow) of fusion plasmas due to excessive heat loads on the probe and its housing structure. Probes also locally perturb the plasma and may affect global plasma parameters which can enhance the probability for disruption. In addition, spurious signals resulting from sheath nonlinearities at the plasma–probe boundary prevent information from being obtained from the true fusion plasma, as well as introducing false signals in the presence of rf fields.

The rapid development of far-infrared (FIR) cw sources and detector technology over the past decade has enabled collective scattering studies from waves in both laboratory and fusion plasmas to be realized in this region of the electromagnetic spectrum. Specifically, cw scattering has been successfully performed from grid-launched ion acoustic waves in an unmagnetized filament discharge plasma (Park et al., 1980, Park, 1980), from density perturbations associated with resonance absorption of electromagnetic radiation in an unmagnetized inhomogeneous plasma (Yu et al., 1983), from optical mixing and stimulated Brillouin scattering generated in ion acoustic waves in a laboratory plasma (Pawley et al., 1985), from spontaneously occurring low-frequency microturbulence in a tokamak plasma (Semet et al., 1980; Taylor et al., 1980), and from mode-converted ion Bernstein waves associated with tokamak ICRF heating (P. Lee et al., 1982a; Park et al., 1984a). Unfortunately, due to the single channel nature of the measurements, wave dispersion relationships were obtained by shot-to-shot variation of the scattering collection angle. Therefore, the reliability of the frequency and wavenumber spectra were completely dependent upon the reproducibility of the plasma and the phenomena under investigation. In the study of random processes such as low-frequency microturbulence, the fact that there is only a single channel represents an obvious restriction. In addition, diagnostic applications such as tokamak ion temperature measurement via cw scattering from externally launched waves are dependent upon single shot time resolved wave dispersion data (Park et al., 1984b).

To satisfy the above mentioned needs, a multiangle, multimixer far-infrared scattering apparatus has been developed at UCLA (Park et al., 1982; Park et al., 1985; Park, 1984; Brower, 1984). In this type of system, the FIR radiation scattered from plasma density fluctuations is collected by six mirrors which are oriented to accept radiation corresponding to six discrete wavenumbers. In the case of a 1200(400) μm wavelength probe beam, this corresponds to obtaining spectral information in the 0–20(60) cm^{-1} region for scattering angles ranging from 0–22°. System flexibility permits considerable variation to larger wavenumbers should the experimental situation dictate the need to do so. The frequency shifted scattered radiation in each channel is combined with a local oscillator beam and downconverted in a quasi-optical Schottky barrier diode mixer (Gustincic, 1977a). Separate low-noise IF amplifier systems are employed to amplify the frequency downconverted signals to a level sufficiently high for subsequent data processing. In order to accurately obtain dispersion relations as well as the absolute level of the density fluctuations, considerable effort has been devoted to system calibration. To achieve this goal, system parameters such as collection efficiency, wavenumber resolution, scattering volume, and detector sensitivities have been carefully measured. The calibration procedures have included use of acoustic cells (Park et al., 1982; Gordon, 1966; Saito et al., 1981; Vogel and Dodel,

1984), rotating wire scattering targets (Dodel et al., 1984), laboratory test wave scattering (Park et al., 1980), as well as standard hot-cold load techniques (Park, 1980).

This chapter presents a thorough description of a multichannel far-infrared scattering system which permits acquisition of spatially and temporally resolved single-shot dispersion data. Section II contains a brief review of collective Thomson scattering theory. The choice of scattering in the far-infrared region of the electromagnetic spectrum will be discussed, along with problems regarding the interpretation of scattering results. A general description of the FIR radiation source will follow in Section III. The design of the optical scattering system and detection techniques are detailed in Section IV. Section V describes the Schottky barrier diode mixers, IF amplifiers, and data acquisition-processing–systems. Perhaps most importantly, the calibration procedures including design of the acoustic cell are presented in Section VI. Typical scattering data from the UCLA Microtor tokamak and the University of Texas TEXT tokamak are shown in Section VII. Finally, through relatively simple system changes, the scattering apparatus can be reconfigured to perform other diagnostic measurements essential to tokamak physics, which will be referred to in Section VIII.

II. Review of Collective Thomson Scattering Theory

In this section a brief review of the salient features of scattering theory is presented, together with a summary of the relevant equations. This will permit an assessment of the sensitivity and resolution of the far-infrared scattering apparatus. For more details the reader is referred to the many excellent references available (Sheffield, 1975; Tsukishima, 1979; Slusher and Surko, 1980).

A. General Scattering Features

An electron moving in the field of an electromagnetic wave experiences an acceleration and emits electric dipole radiation. The form of the spectrum of electromagnetic radiation scattered by a charged particle depends on its mass, charge, velocity, and position. Similarly, the spectrum of radiation scattered by a plasma, an assembly of charges, depends on the plasma properties. Consider a plane wave

$$\tilde{E} = \hat{z}E_0 \cos (k_0 z - \omega_0 t) \tag{1}$$

incident on a plasma as shown in Fig. 1. The superposition of the scattered radiation from a collection of electrons that are distributed com-

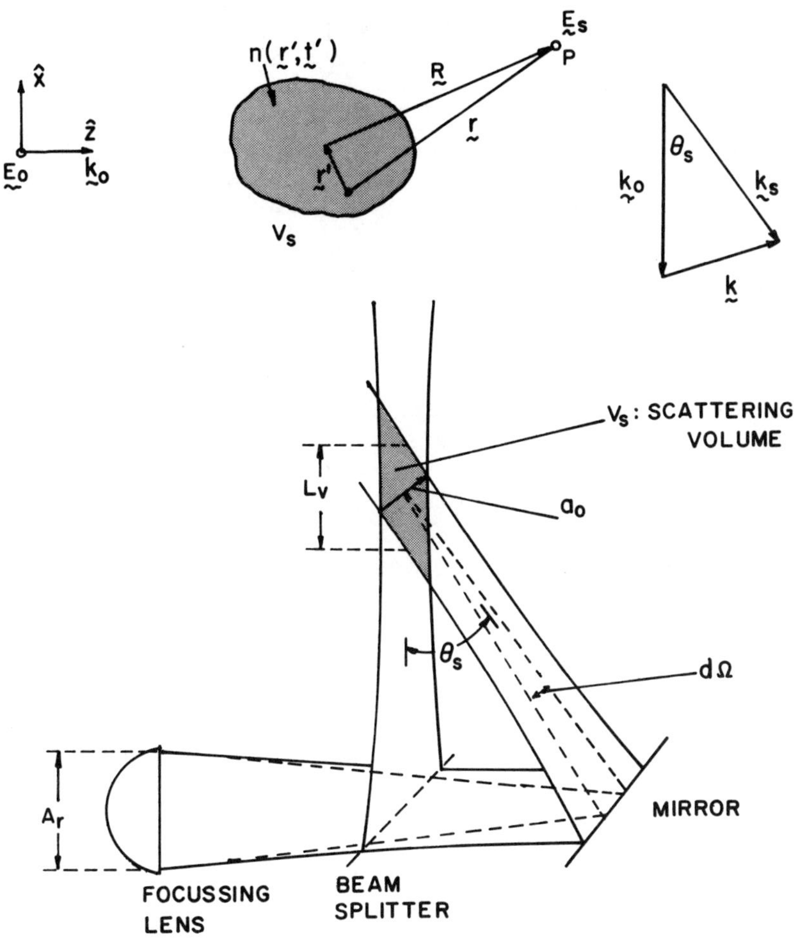

FIG. 1 Scattering geometry.

pletely uniformly interferes destructively and no net scattering results. However, deviations in density occur due to random thermal fluctuations and also due to coherent collective motions in a plasma. These departures from uniformity give rise to a scattered intensity proportional to the density fluctuation amplitude, $\tilde{n}$, for random fluctuations and to $(\tilde{n})^2$ for coherent fluctuations. The scattered power is inversely proportional to the square of the mass of the charges; therefore, scattering from electrons dominates.

The reradiated power from the accelerated electron occurs at a frequency $\omega_s = \omega_0 + (\mathbf{k}_s - \mathbf{k}_o) \cdot \mathbf{v}_e$ where $\mathbf{v}_e$ is the electron velocity. If the fluctuations have a frequency ω and wavenumber $\mathbf{k}$, the scattered wave $(\omega_s\ \mathbf{k}_s)$ must satisfy momentum and energy conservation so that

$$\omega_s = \omega_0 \pm \omega \qquad \text{and} \quad \mathbf{k}_s = \mathbf{k}_o \pm \mathbf{k}, \tag{2}$$

where $(\omega_0, \mathbf{k}_o)$ describe the probe beam. For cases of interest, $|\mathbf{k}_o| \simeq |\mathbf{k}_s|$, so that $k^2 \simeq 4k_o^2 \sin^2(\theta_s/2)$ where θ_s is the scattering angle. This yields

$$\theta_s = 2 \sin^{-1}\left(\frac{k}{2k_o}\right), \tag{3}$$

the familiar Bragg relation.

Collective Thomson scattering is differentiated from ordinary Thomson scattering by examining the electromagnetic wave interaction with charges on a scale larger than a Debye length $[\lambda_D = (k_B T_e/4\pi n_e e^2)^{1/2}]$, where k_B, T_e, n_e are Boltzmann's constant, electron temperature, and electron density, respectively. It is useful to define a characteristic parameter, α, that is essentially the ratio of the effective scattering wavelength to the Debye length:

$$\alpha = \frac{1}{k\lambda_D} \simeq \frac{1}{2k_o\lambda_D \sin\left(\frac{\theta_s}{2}\right)} = \frac{\lambda_o}{4\pi\lambda_D \sin\left(\frac{\theta_s}{2}\right)}. \tag{4}$$

For $\alpha \ll 1$, scattering is from individual (uncorrelated) electrons and the scattered radiation has a frequency spread determined by the electron temperature. In the $\alpha \geq 1$ limit, information is obtained on ion fluctuations and, therefore, the ion temperature in a thermal plasma, through scattering from the electrons whose motion is correlated with the ions. Alternatively, in the case of a nonthermal plasma, enhanced reradiation may result due to coherent scattering from a density wave. Before examining in detail the implications of $\alpha \geq 1$ for scattering, let us first complete our brief review of scattering theory.

The scattered power P_s is related to the incident power P_o by

$$P_s = P_o r_e^2 n_e L_v S(\mathbf{k}) \Delta\Omega_r, \tag{5}$$

where n_e is the plasma density, L_v is the scattering volume (V_s) interaction length (see Fig. 1), $\Delta\Omega_r$ is the collection solid angle, and r_e is the classical electron radius. The spectral density function or scattering form factor $S(\mathbf{k})$ is given by

$$S(\mathbf{k}) = \frac{n_e}{TV_s} \int_{-\infty}^{+\infty} \left|\frac{\tilde{n}(\mathbf{k},\omega)}{n_e}\right|^2 \frac{d\omega}{2\pi} = \int S(\mathbf{k},\omega) \frac{d\omega}{2\pi}, \tag{6}$$

where T is the time over which the measurement is made. In the case of a coherent density fluctuation [i.e., $n = \tilde{n} \cos (kx - \omega t)$] one obtains

$$P_s = \frac{1}{4} P_o r_e^2 \lambda_o^2 (\tilde{n})^2 L_v^2. \tag{7}$$

The scattering will have an angular extent, $\Delta\theta$, related to the incident laser beam radius, a_o (measured at the e^{-2} power points), given by

$$\Delta\theta = \frac{\lambda_o}{\pi a_o}, \tag{8}$$

which is the half angle divergence of a Gaussian beam. For small angle scattering, Eq. (8) and the Bragg condition combine to yield the wavenumber resolution

$$\Delta k \simeq k\Delta\theta = \frac{2}{a_o} \tag{9}$$

which is independent of k. In the case of a broad spectrum of waves, $S(\mathbf{k},\omega)$ can be obtained experimentally with a multichannel system. Then, using Eq. (6), the absolute value of the density fluctuations can be obtained through

$$\left|\frac{\tilde{n}(t)}{n_e}\right|^2 = \frac{1}{n_e} \int S(\mathbf{k}) \frac{d\mathbf{k}}{(2\pi)^3} \tag{10}$$

The spatial resolution along the direction of $\mathbf{k}_o$ is estimated by

$$L_v = \frac{2a_o}{\sin \theta_s} \tag{11}$$

The spatial resolution perpendicular to $\mathbf{k}_o$ is simply $\pm a_o$.

Implications of the above relations dictate the optimal choice of scattering wavelength (i.e. microwave, far-infrared, infrared) for a particular plasma. The first requirement is that $\alpha \geq 1$ for collective scattering; i.e. for fixed λ_D, λ_o must be large or θ_s small. For infrared scattering (i.e. CO_2 laser) θ_s is typically less than one degree, which results in poor spatial resolution (i.e. large L_v) and prohibits a multiangle measurement. Slusher and Surko (1978) have developed cross beam correlation techniques which significantly improve spatial resolution for CO_2 scattering systems. Unfortunately, simultaneous achievement of good spatial and wavenumber resolution does not appear feasible with such a system. These constraints therefore argue for a microwave or far-infrared scattering system. However, several new problems arise at longer wavelengths. First, microwave probe beams can suffer significant refraction or even cutoff for typical tokamak fusion plasmas. Refraction affects both the incident and scattered beams, thereby necessitating corrections to the wavenumber and the scattering volume being probed. For large refraction effects, such corrections may contribute substantially to the experimental error. Second, the plasma tends to be optically thick to the first few cyclotron harmonics, resulting in a significant synchrotron radiation background at longer probe wavelengths for thermal plasmas. In addition, a small population of runaway electrons will considerably enhance the radiation at the cyclotron frequency and its harmonics. Finally, for many devices it is required to propagate both the probe and scattered beams over long path lengths without guiding structures. From Eq. (8), the diffraction losses are obviously more severe at longer wavelengths since a_o is limited by port access.

The choice of a far-infrared probe beam results in a good compromise between the various criteria mentioned above. This technique is capable of providing a nonperturbing determination of the ω and $\mathbf{k}$ spectra of nonthermal density fluctuations for plasmas with densities in the range of 10^8–10^{16} cm^{-3}. The flexibility in choice of far-infrared wavelength (100–2000 μm) allows refraction effects to be minimized while still permitting collective scattering at the large angles necessary to realize a multichannel system. In addition, the use of two probe wavelengths enables the same $\mathbf{k}$ to be observed at different scattering angles, therefore, by using separately calibrated systems one can cross-check the spectra and absolute fluctuation levels. However, large scattering angles do require substantial access to the plasma via vacuum windows, a luxury not available on all devices.

B. Interpretation of Scattering Results

Collective Thomson scattering is widely recognized as a nonperturbing diagnostic for the study of density fluctuations in both laboratory and fusion plasmas. However, interpretation of data is not always trivial. For example, great care must be exercised in assessing the wavenumber matching conditions before attempting to deduce absolute values for density fluctuation amplitudes and wavenumber spectra.

Consider the two-dimensional geometry illustrated in Fig. 1. An electromagnetic wave is incident in the $\hat{z}$ direction ($\mathbf{k}_0$) and scattering is observed in a direction $\mathbf{k}_s$ at some angle θ_s to the incident beam. Therefore, $\mathbf{k}_+$, the scattering wave vector, is given by Eq. (2). We will assume a Gaussian incident beam profile and density fluctuation given by

$$n_e(r', t') = \tilde{n} \cos (k_x x' + k_z z' - \omega t'). \tag{12}$$

In addition, it is assumed that the area where the fluctuation exists is much larger than the probe beam waist and finite along the beam with arbitrary length L. The scattered power can be shown to be

$$P_s = \frac{1}{4} P_o r_e^2 \lambda_o^2 \tilde{n}^2 L^2 F(\mathbf{k}_+, \mathbf{k}) \tag{13}$$

for a given solid angle where

$$F(\mathbf{k}_+, \mathbf{k}) = \exp\left[-\frac{a_o^2}{2}(k_x - k_{+x})^2\right] \operatorname{sinc}^2\left[(k_z - k_{+z})\frac{L}{2}\right] \tag{14}$$

and

$$\operatorname{sinc}(x) = \frac{\sin(x)}{x}$$

When L is larger than the intrinsic scattering length L_v defined by Eq. (11), then L can be replaced with L_v. In this case, the sinc function contribution is negligible. On the other hand, when L is smaller than L_v, L cannot be replaced with L_v and the sinc function contribution is significant.

For the case of perfect matching ($k_{+x} = k_x$ and $k_{+z} = k_z$) one obtains the familiar result given in Eq. (7). However, let us consider the more general situation of a wavenumber mismatch. Consider Fig. 2(a), where a density wave is propagating in the matched direction but is not matched in amplitude, i.e. $|\mathbf{k}_+| \neq |\mathbf{k}|$. In this situation, the Gaussian term dominates over the sinc function and the often quoted k-resolution of the scattering system is described. A variation of wavenumber produces a Gaussian falloff in the scattered power with resolution $\Delta k = 2/a_o$.

Now let us assume an angular mismatch exists. In this case the wavenumber amplitude is matched ($|\mathbf{k}_+| = |\mathbf{k}|$) but the wave is propagating in a mismatched direction. This is illustrated in Fig. 2(b). Under these conditions the dominant term in Eq. (13), is the sinc function and not the Gaussian. This means that as the wave direction is varied the scattered power decays as a sinc function whose width becomes narrower at larger $|\mathbf{k}|$.

In order to verify the above experimentally, an acoustic cell which will be thoroughly discussed in Section V was employed. A far-infrared laser beam is scattered from unidirectional waves produced in a piezoelectrically driven TPX acoustic cell. The cell can be rotated to angularly match or mismatch the scattering system. In the matched direction the wavenumber amplitude is easily varied, changing the frequency of the driven wave. This results in the expected Gaussian variation in scattered power and a measured k-resolution of $\simeq 1\ \text{cm}^{-1}$, which agrees well with

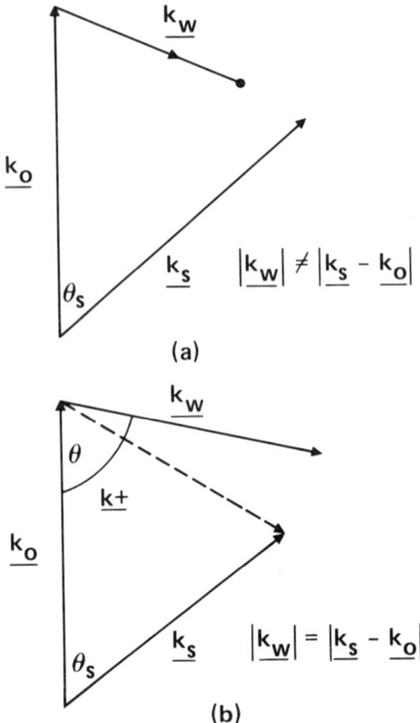

FIG. 2 Diagram of wavenumber mismatch; (a) magnitude mismatch, (b) angular mismatch.

the theoretical value of $2/a_o$ discussed previously. These results will be given in more detail in Section VI.

Figure 3 illustrates the results obtained when, for a particular scattering angle with $|\mathbf{k}_+| = |\mathbf{k}|$, the acoustic cell is rotated to angularly mismatch the scattering system. This was repeated for a number of scattering angles and equivalent $|\mathbf{k}|$. As can be seen, the experimental results agree quite well with theoretical sinc function curves. In particular, the narrowing of the angular acceptance range at higher $|\mathbf{k}|$ is clearly demonstrated. It should be noted that, since the sinc function narrows at larger wavenumber, any applied correction will also vary with wavenumber. For example, at long wavelengths the corrections are slight whereas at shorter wavelengths the opposite is true. For the Microtor tokamak plasma (L = 20 cm), as shown in the theoretical results of Eq. (14) in Fig. 4, corrections for the power spectrum at different wavenumbers have been reduced compared to the acoustic cell case (L = 5 cm).

Another important problem in interpretation of the measured scattering amplitude is the curvature of the target wave phase front along the inci-

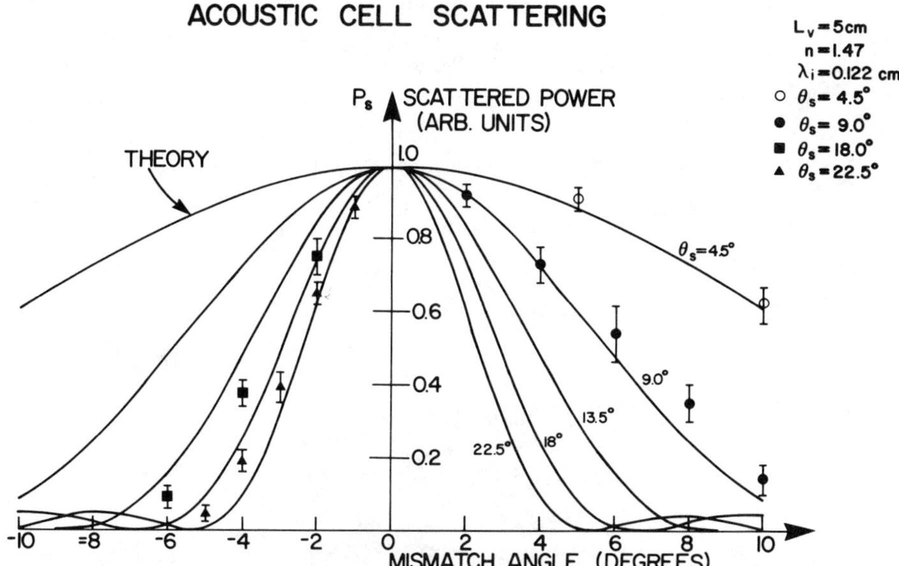

FIG. 3 Experimental results of scattered power as a function of the mismatch angle.

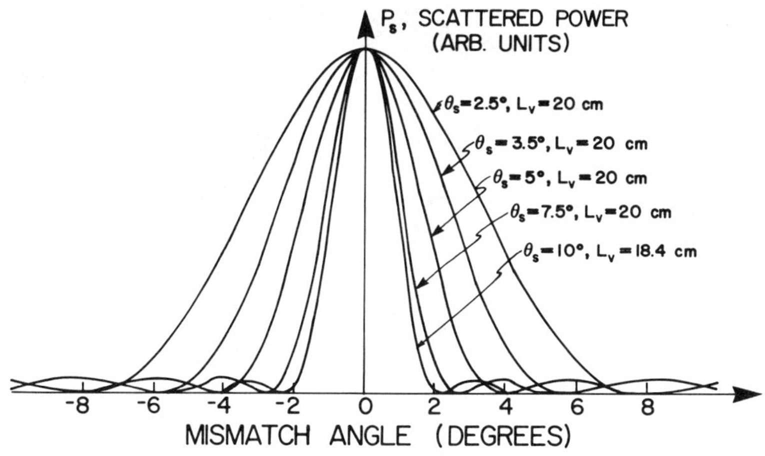

FIG. 4 Theoretical results of the scattered power as a function of the mismatch angle for the Microtor tokamak.

dent probe beam. The scattered radiation can be combined constructively or destructively depending upon the degree of wave front curvature. This effect is dramatically demonstrated in Fig. 5. Here, two acoustic cells possessing the same frequency wave packet are employed. These cells are physically arranged to produce constructive or destructive interference. This implies that one could apparently observe no wave or a wave

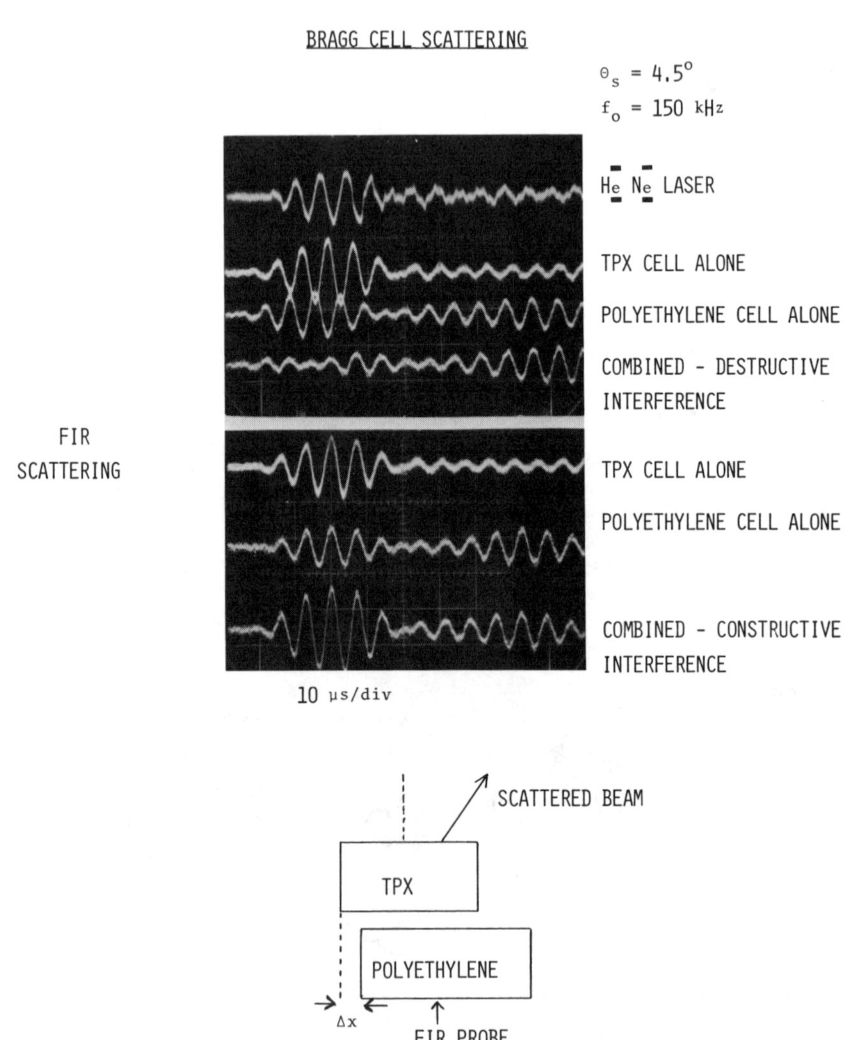

FIG. 5 Demonstration of constructive and destructive interference in the scattering signal produced by the use of two acoustic cells.

with twice the driven value, even though there is the same amplitude wave in both cases.

The impact of the above results on scattering experiments performed in fusion plasmas, where coherent waves are being studied, can be simply illustrated in a laboratory plasma. Radiation from a 1222 μm laser is used to scatter from mesh launched ion acoustic waves in an unmagnetized filament discharge plasma ($n_e \simeq 1 \times 10^{10}$ cm^{-3}, $T_e \simeq 3$ eV, $T_e/T_i \simeq 8$). Initially, a 15 cm square, flat mesh was used to launch unidirectional ion waves. The mesh was rotatable to allow the wave direction to be matched or mismatched. Figure 6 illustrates the experimental results. The scattered power was found to vary with mismatch angle as a sinc function. However, the use of a cylindrical mesh to launch waves in a variety of directions resulted in the peak scattered power being reduced and to be virtually constant with scattering angle. Locally, the wave amplitude was maintained equal for both the flat and the cylindrical mesh, as monitored with a Langmuir probe. The variation in the wave propagation direction as a function of position has resulted in only waves restricted to a small spatial region being detected. Clearly, an accurate deduction of the density fluctuation amplitude is dependent upon an a priori knowledge of the wave directionality as a function of position along the scattering length.

An estimation of the absolute density fluctuation amplitude via collective scattering may only be obtained simply if the density fluctuation is homogeneous within the scattering volume and the wave directionality is well known. However, in the case of actual fusion plasma applications, the information concerning wave fronts, amplitude distribution, etc. is not generally available. Particularly for small angle scattering, sometimes the intrinsic scattering volume length L_v is much larger than the actual device size. Therefore, the analytical expressions for the scattered power and electric field cannot be applied without considering the associated problems. For instance, in the study of coherent waves, one should consider the phase interference difficulties along the scattering length. Even in the case of random isotropic density fluctuation studies such as microturbulence, in order to obtain the wavenumber spectra, one should be aware of the possible corrections due to wavenumber mismatch problems arising from the short interaction length.

As a final point, we note that wave damping or dissipation can modify the angular distribution of the scattered radiation. Specifically, if the wave amplitude decreases as the wave propagates across the scattering volume, the angular distribution of the scattered radiation is broadened even if the wavelength remains constant. This potential complication can also be viewed as an aid in the sense that it allows one to assess wave damping. This would permit one to perform such important measurements as the

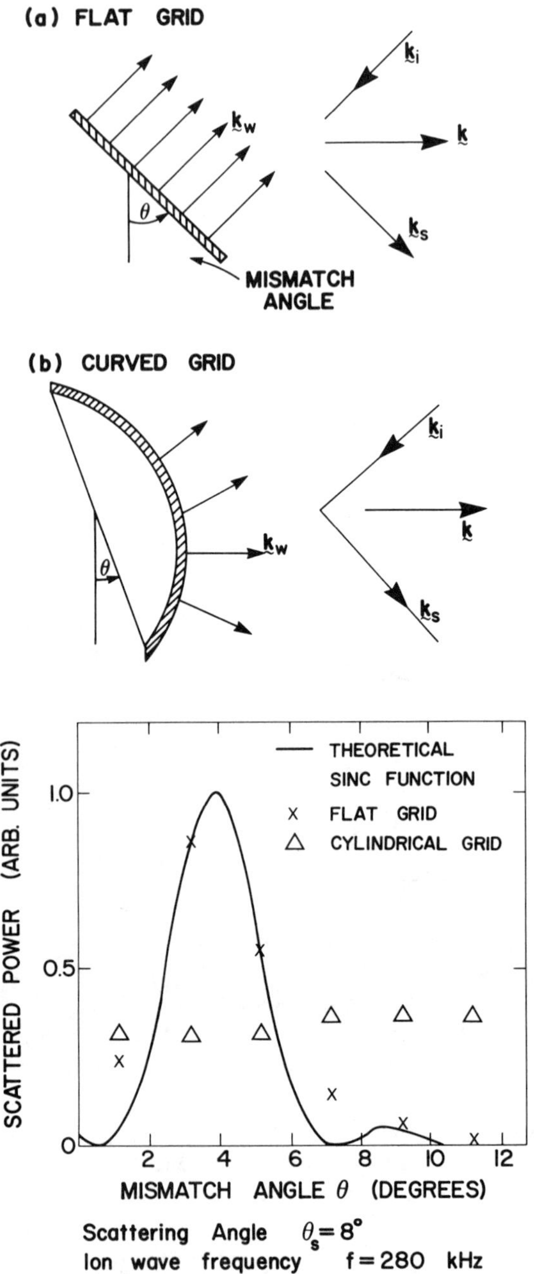

FIG. 6 Scattered power from ion waves launched with flat and curved grids as a function of the mismatch angle.

determination of the α-particle distribution in fusion plasmas by the selective excitation of waves which are damped by the α-particles. In a preliminary series of experiments, this appears to be observed in the case of ion acoustic waves generated via optical mixing where the increased damping is due to ion tail formation (Pawley, 1985).

III. Far-Infrared Probe Source

The far-infrared source employed in the multichannel scattering experiments is an optically pumped molecular gas laser with discrete lines available in the 100–2000 μm range. Other sources including electrical discharge lasers (HCN, 337 μm; Kneubuhl and Sturzenegger, 1980), GUNN diodes ($\leq$2 mm, Crowley, 1983), klystrons ($<$2 mm), and high frequency backward wave oscillators called carcinotrons ($\geq$500 μm; Kantorowicz and Palluel, 1979; Palluel and Goldberger, 1956) may also be used. Lasers are generally preferred as they provide narrow bandwidth radiation with sufficient power output to perform scattering measurements with a favorable signal-to-noise ratio.

The cw FIR laser, to be discussed in this section, consists of a CO_2 "pump" laser whose output is absorbed by a molecular gas contained in an FIR laser cavity, as shown in Fig. 7. The optical pumping process is relatively simple, as indicated schematically in Fig. 8. Here, the near coincidence of the vibrational-rotational absorption line of a polar molecule and an infrared laser (i.e. CO_2) emission line results in the selective population of a particular rotational sublevel in an excited vibrational state. The far-infrared lasing transition occurs between adjacent rotational states in the upper vibrational manifold. The FIR laser design is similar to that developed by Hodges et al. (1976) except that the FIR cavity is optimized for long wavelength (400–1200 μm) operation.

A. CO_2 Pump Laser

For the experiments described herein, the CO_2 laser discharge tube is shown schematically in Fig. 7 and consists of a self-aperturing waveguide with an inner diameter of 8.5 mm and a length of 160 cm. ZnSe Brewster windows are attached to the discharge tube by an O ring seal. The CO_2 cavity consists of a ZnSe output coupler (f.l.:   10 m concave; reflectivity: 50%) and a ruled grating (53 lines/cm). A piezoelectric stabilizer is mounted on the output coupler providing fine control over the cavity length for tuning of the infrared frequency. For thermal stability, the entire discharge tube is water cooled (15°C) and mounted on Invar support rods, electrically isolated by Teflon sleeves. The support system

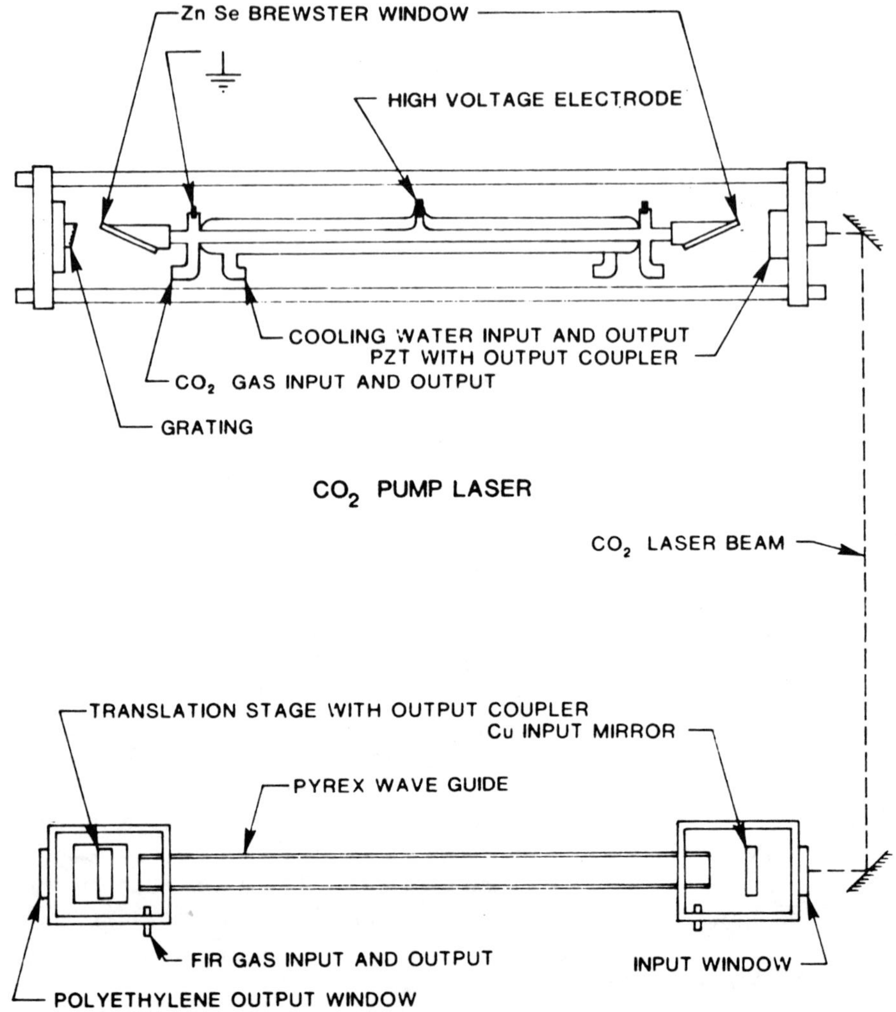

FIG. 7 Schematic of the optically pumped far-infrared (FIR) laser.

including the grating mount is also water cooled to improve temperature stability. The optimum gas mixture of 13% CO_2, 22% N_2, and 65% He, is caused to flow through the system. A gas pressure of 60 Torr provided the maximum power output for typical discharge voltages and currents of 13

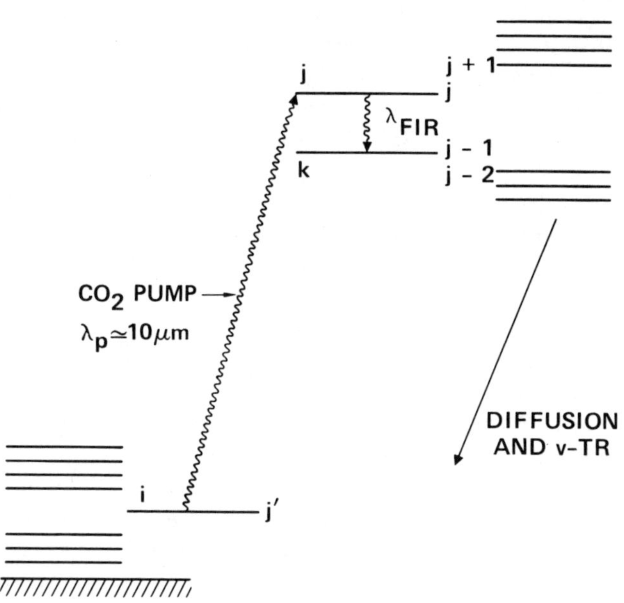

FIG. 8 Schematic energy level diagram of a polar molecule illustrating the optical pumping process (courtesy D. T. Hodges).

kV and 25 mA. The discharge tube has a high voltage electrode at the center and both ends are grounded as shown in Fig. 7.

The CO_2 laser output power was $\simeq 50$ W when the grating was tuned for operation on the $9P(32)$ line at 9.6 μm (used for 1222 μm FIR laser); however, recent advances in laser design (Peebles, 1984) have been able to produce up to 150 W. The entire system is sufficiently stabilized to operate indefinitely. Heat deposition on the optical elements only affects operation if the components are "dirty", thus considerably increasing absorption. Multiple transverse modes are eliminated by inserting a stainless steel ring in the discharge tube, which effectively reduces the aperture and decreases the gain for higher order modes.

Although the CO_2 laser described above produced sufficient pump strength to provide the FIR output power required for mixer bias and probe power, there has recently been undertaken a collaborative high-power CO_2 FIR laser development project between UCLA and Apollo Lasers (Peebles, 1984; Apollo Lasers, Inc., 1984). A higher-power laser is required in order to increase signal-to-noise ratios and to enable the scattered spectra to be observed at larger wavenumbers. Here it should be noted that even relatively small increases in the FIR output power can result in dramatic increases in the signal-to-noise ratio, as can be simply demonstrated. For example, the FIR laser which has been employed to date produces $\simeq$ 8 mW of stable output at 1222 μm. Approximately 6 mW of this is used to rf bias the six Schottky diode mixers, leaving $\simeq$ 1–2 mW for the probe beam. The newly developed CO_2 FIR laser package will produce $\simeq$ 30–40 mW at 1222 μm. Again 6 mW will be required for the local oscillator power leaving $\simeq$ 25–35 mW for the probe beam. This represents an increase in scattered power and S/N $\simeq$ 20. Such an increase will significantly improve the quality of the data and also enables one to probe fluctuations with larger wavenumber. Low-level fluctuations are observed at increased scattering angles ($\simeq$20°) where the spatial resolution is considerably improved ($\simeq$5 cm).

To increase the FIR output power the first step is to redesign the CO_2 laser as mentioned above. Figure 9 shows a schematic representation of the new CO_2 pump laser. The unique features include a four electrode design which eliminates the need for Brewster windows (resulting in increased stability) as well as the expense of laborious glass blowing, etc. As shown in Figs. 10(a) and 10(b), output power levels in excess of 150 W (a factor of 3–4 improvement over the previous design) are obtainable at most wavelengths in the 9–11 μm region with a maximum output of $\simeq$200 W having been observed ($\simeq$250 W with a plane mirror replacing the grating in the cavity).

B. Optically Pumped FIR Laser

The FIR laser cavity, employed for the majority of the scattering measurements to date (see Fig. 7), consists of a 5 cm diameter, 180 cm long Pyrex tube dielectric waveguide to which endboxes containing the input-output–couplers are attached. The waveguide is chosen to minimize propagation losses ($\simeq\lambda_0^2/a^3$) and maximize gain ($\simeq\lambda_0/a$) for operation in the 400–1200 μm range (a = waveguide radius). This geometrical arrangement assists in selectively exciting the lowest order mode, EH_{11}, for the dielectric waveguide. The mode is linearly polarized and transforms efficiently into the TEM_{00} Gaussian free space mode. Figure 11 compares the attenuation coefficient as a function of wavelength of the EH_{11} mode with other commonly employed low-loss waveguides.

The CO_2 pump beam is coupled into the FIR laser cavity through a 3–4 mm diameter hole in a plane Cu mirror. The infrared radiation is focussed by a ZnSe lens (f.l.: 40 cm) and enters the vacuum box through a ZnSe Brewster window. The CO_2 beam expands in the FIR cavity and is absorbed by the molecular gas.

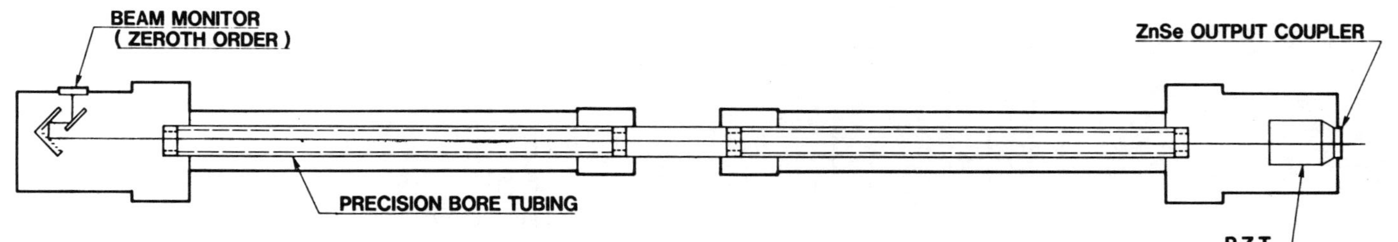

FIG. 9 Schematic of the new CO_2 pump laser designed through UCLA and Apollo Laser, Inc. collaboration.

CO2 Laser Power
170
0
CO2 Output (Watts)
R42
R4
P6
P52
9
10
Wavelength (Microns)

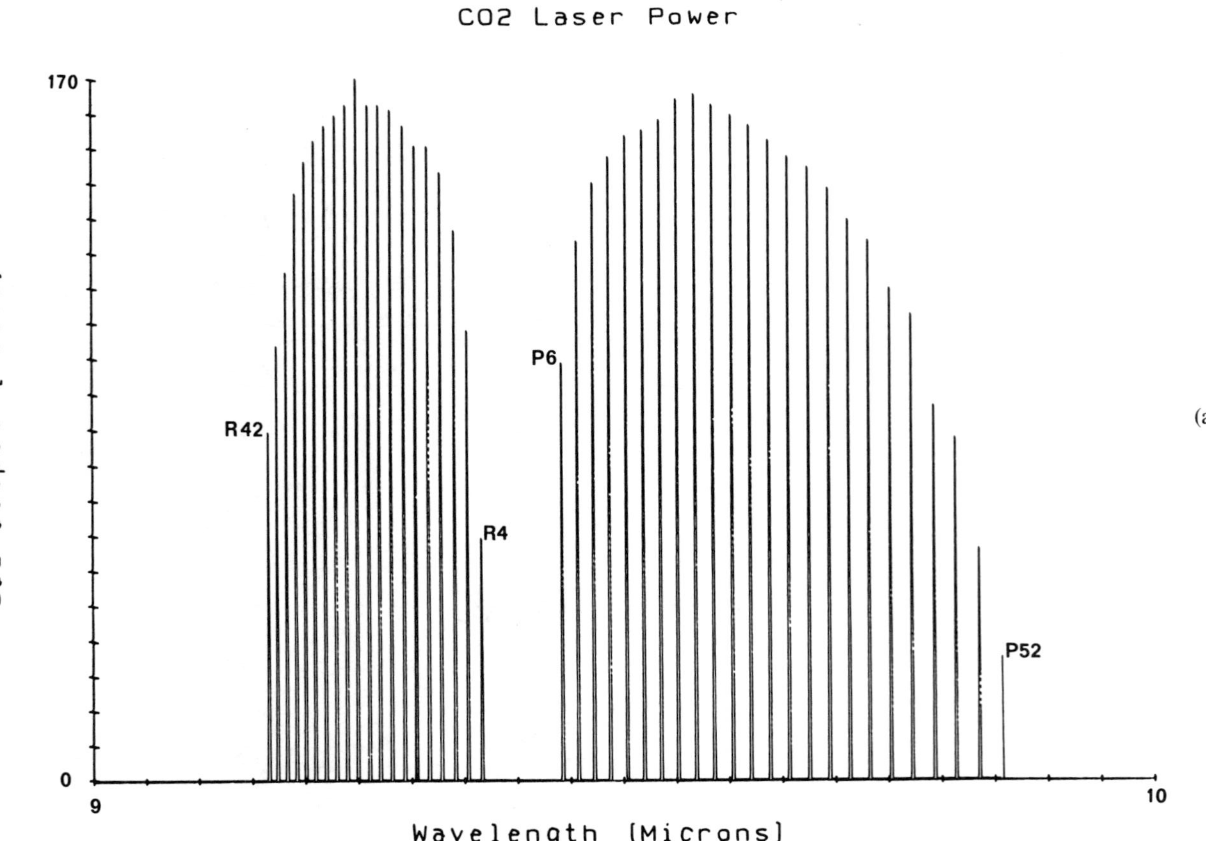

(a)

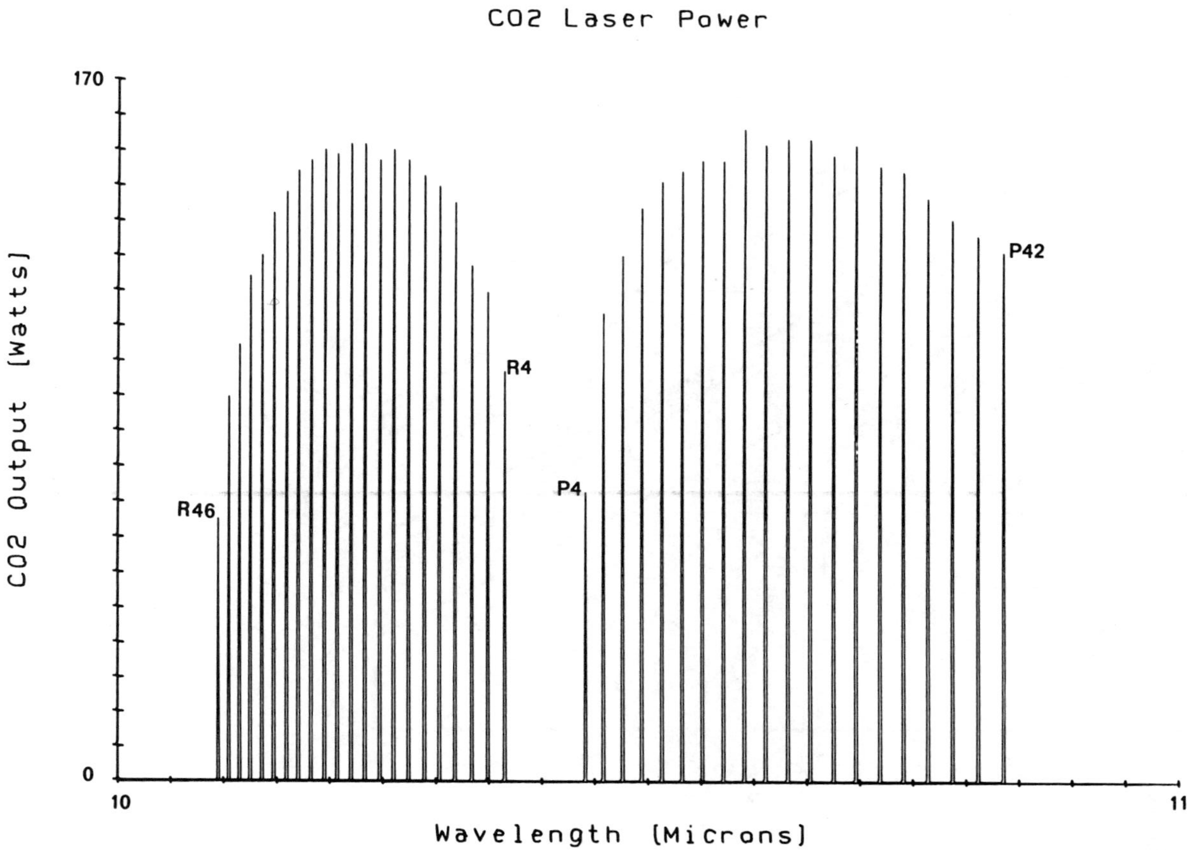

CO2 Laser Power

CO2 Output [Watts]

170

0

10

11

R46

R4

P4

P42

Wavelength [Microns]

(b)

FIG. 10 Output power of the infrared pump laser at various lines in the (a) 9 μm branch, and (b) 10 μm branch.

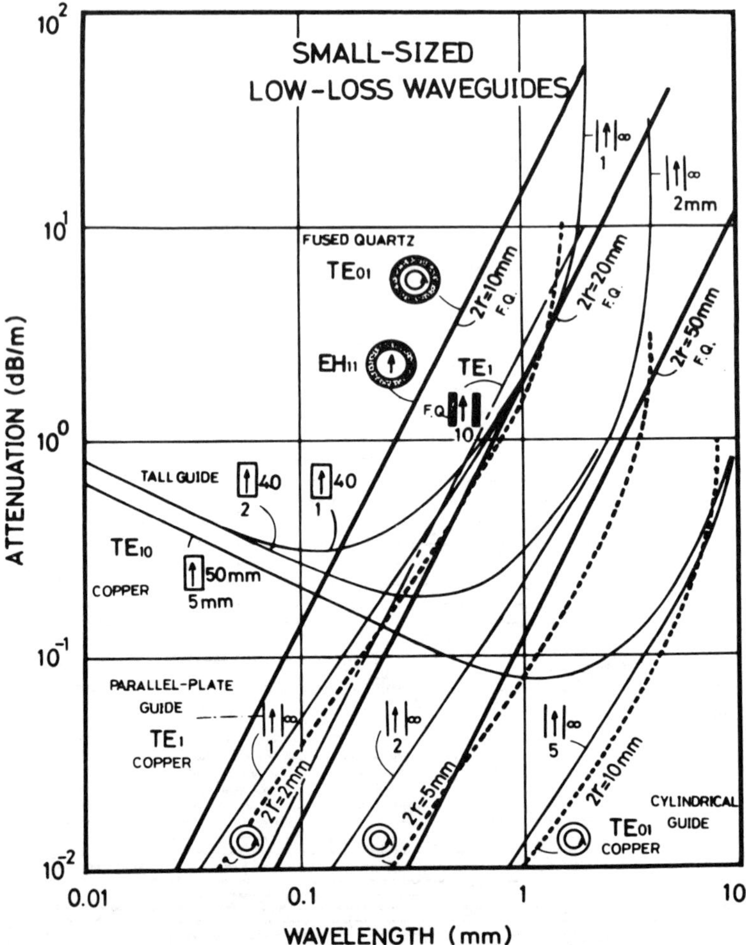

FIG. 11 Attenuation coefficient as a function of wavelength for both copper and quartz wavelengths (Yamanaka, 1977).

An inductive mesh serves as the FIR output coupler. The mesh is chosen to provide optimal far-infrared output power for the pump source, cavity length, molecular gas pressure, and wavelength designed for any particular system. A reflection coated *z*-cut quartz crystal is located directly in front of the mesh and feeds the infrared pump radiation, which has not been absorbed, back through the molecular gas medium. The quartz represents negligible attenuation to the FIR while the thin reflection coating is only seen by the infrared. The FIR cavity is operated at a

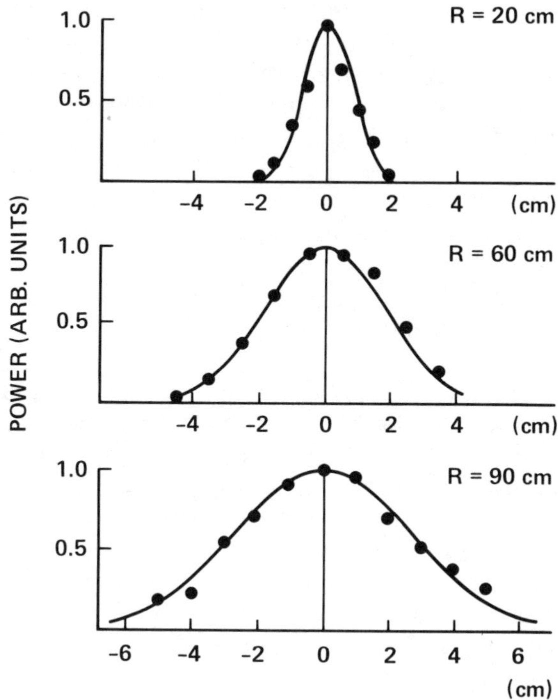

FIG. 12 FIR laser beam profiles as a function of distance from the cavity output window.

pressure of 50 mTorr for a $C^{13}H_3F$ molecular gas producing output at 1222 μm.

The output of the FIR laser is characterized by a Gaussian TEM_{00} mode as determined by beam profile measurements shown in Fig. 12. A pyroelectric detector with 2 mm aperture was scanned across the beam at various distances from the output coupler. The FIR laser beam was chopped at about 100 Hz. This figure shows the beam intensity profiles at distances of 20, 60, and 92 cm from the output coupler with corresponding beam widths (e^{-2} point) of 1.5, 3.9, and 5.4 cm, respectively. The profiles are observed to make good fits to a Gaussian lineshape.

All FIR power measurements are performed with a model 3620 Scientech thermopile power meter. Foote and Hodges (1979) have estimated the absorption of the Scientech thermopile element by measuring the transmitted and reflected portions of an incident FIR beam at various wavelengths. The power calibration factors are 1.3 ± 0.2 and 1.9 ± 0.2 at wavelengths of 1222 and 447 μm, respectively. The typical corrected laser output powers are 10 mW at 1222 μm and 50 mW at 447 μm (CH_3I).

The above results are improved by a factor of 3–4 with the new high-power CO_2 pump laser designed by UCLA and Apollo Lasers. Such a system is presently being tested for replacement of the current FIR laser on the TEXT multichannel scattering system. However, implementation of the new infrared pump laser also requires that changes to the FIR laser cavity be made. For example, the molecular gas temperature (which increases with pump power) must be optimized to provide the appropriate energy level population distribution and consequently the optimum output power. This is illustrated for the case of the 393 μm HCOOH line in Fig. 13. The dramatic power increase at the optimum temperature obviously justifies the added expense of including a cooling system on the FIR cavity.

The considerable increase in CO_2 pump power has resulted in other interesting differences in operating behaviour as compared to FIR lasers pumped by lower-power CO^2 lasers (i.e. $\simeq$30–50 W). For example, it is well known that in low-power $C^{12}H_3F$ 496 μm lasers modest power and efficiency increases (i.e. $\simeq$50%) can be obtained by the addition of Hexane (Hodges et al., 1976). However, the behaviour is entirely different at high power. Here, we find the power output to strongly saturate in the absence of hexane as expected due to the "vibrational bottleneck" characteristic of such diffusion limited systems (DeTemple and Danielewicz, 1983). Hence, in this case the addition of hexane results in a threefold increase in output power (T. Lehecka, private communication, 1985).

The uncorrected output power (Scientech power meter) levels for various lines are shown in Table 1. As mentioned previously, the FIR laser

TABLE I

SUMMARY OF FIR LASER PERFORMANCE ON SELECTED LINES

λ_{FIR} (Microns)	Power (mW)	Gas	λ_{pump}	Pump power (watts)
185	400	CH_2F_2	9.21 R32	140
393	140	HCOOH	9.28 R18	150
432	65	HCOOH	9.27 R20	150
496	80	CH_3F	9 P20	140
743	25	HCOOH	9.17 R40	90
1222	35	$C^{13}H_3F$	9.66 P32	140

must satisfy several other design constraints in addition to power output. For example, the long propagation distances and imaging optics required mandate the use of a Gaussian beam. As shown in Fig. 14, the new FIR laser does indeed produce an EH_{11} mode which efficiently transforms into a Gaussian free space mode. In addition, narrowline output and stability are extremely important. However, as shown in Fig. 15, the new FIR laser is extremely narrowline and stable.

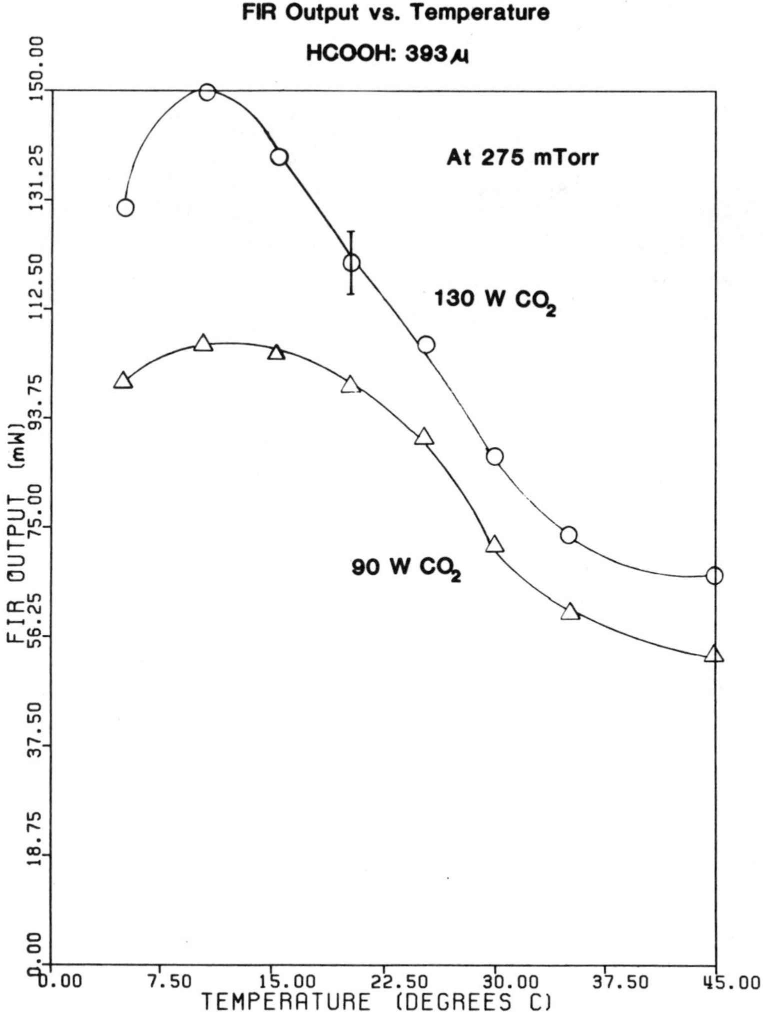

FIG. 13 Output of HCOOH (393 μm) laser as a function of temperature.

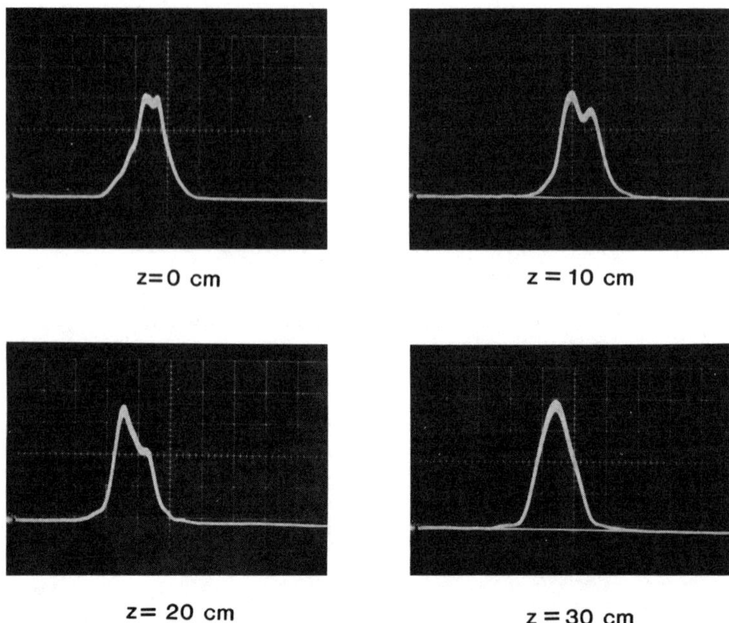

FIG. 14 Evolution of the EH_{11} mode in the near field (experimental).

C. OTHER FIR SOURCES

Electrically excited far-infrared lasers provide coherent output at discrete wavelengths from 28–774 μm. A comprehensive review of electrically excited submillimeter wave lasers is given by Kneubhl and Sturzenegger (1980). For applications to magnetic fusion plasma diagnostics there are three electrical discharge lasers of particular interest. These are the DCN laser which produces $\simeq$250 mW at 195 μm (Veron et al., 1978; Belland and Veron, 1980), the H_2O laser from which $\simeq$50 mW has been obtained at 118.6 μm (Belland, 1982), and also the HCN laser which can produce up to $\simeq$170 mW at 337 μm (Belland et al., 1976; Belland and Crenn, 1979).

Another viable source for use in far-infrared scattering systems is the carcinotron or O-type backward wave oscillator (Kantorowicz and Palluel, 1979; Palluel and Goldberger, 1956). A variety of tubes are available on a commercial basis from Thomson, CSF. For example, the $CO_o10.1$ device produces nearly a watt at $\simeq$300 GHz. Tubes (TH4218C) also exist in the 360–406 GHz region ($\lambda \simeq$ 800 μm) and produce up to 100 mW.

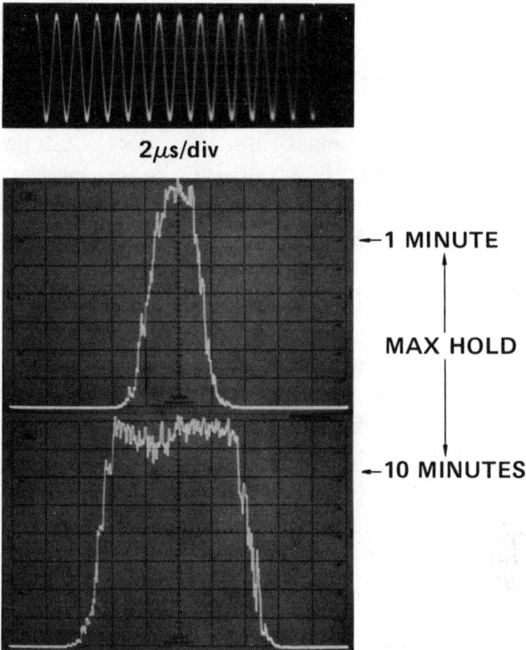

FIG. 15 Spectral output of the "new" FIR laser.

Under a development contract with the European Space Agency Laboratory (ESTEC), Thomson has developed the TH4211 tube which has provided power levels in excess of 50 mW at frequencies above 500 GHz (Thomson, CSF; Th. DeGrauuw, private communication). A problem with the carcinotron is that it is a voltage tunable device ($\simeq$10–20 MHz/V). To provide a narrowline source for studies such as drift wave microturbulence one requires an extremely well regulated low-ripple high-voltage ($\simeq$mW at 10–12 kV) power supply. This requirement, coupled with the need to install sophisticated protective circuitry, has rather limited the power supply market. Up to $\simeq$200 GHz klystrons and extended interaction oscillators produce sufficient power and narrow linewidth to serve as the scattering source. Other tube sources such as gyrotrons (Hirschfield, 1980; Symons and Jory, 1981) produce considerably more power and have been suggested as scattering sources (Biegel and White, 1980; Woskoboinikow et al., 1982). However, thus far no experiments have been performed, therefore leaving open such questions as linewidth, etc.

Solid-state technology will eventually provide suitable sources for millimeter wave scattering systems (local oscillator and probe beams). Currently, GUNN diodes produce up to $\simeq$40 mW at frequencies up to 90 GHz making them useful for lower density plasmas ($\leq 10^{12}$ cm^{-3}). For example, Pawley et al. (1985) have performed detailed scattering studies of the nonlinear saturation of ion acoustic waves generated by the optical mixing of two high-power microwave beams using a phase-locked GUNN homodyne scattering system. IMPATT diodes have provided outputs of $\simeq$25–50 mW at frequencies as high as 240 GHz (Mizuno et al., 1979). Furthermore, 2 mW output has been reported at 400 GHz using LN_2 cooling (Ishibashi et al., 1977). Although IMPATT diodes are rather noisy compared to klystrons, the output can be narrowed using either injection locking (fundamental or subharmonic) or special filtering. A related device called the TUNNETT offers the promise of higher-frequency operation (up to $\simeq$800 GHz) with lower noise albeit with lower output power ($\simeq$1 mW) (Pan and Lee, 1981).

The above mentioned improvements in solid-state technology (Kuno, 1981) have spawned considerable interest in the development of frequency multipliers which combine a nonlinear element (i.e. a varactor) with a high-power lower-frequency solid-state source to generate harmonics at millimeter and submillimeter wavelengths (Danielewicz, 1980; Archer, 1984a, 1984b; Archer and Faber, 1985; Archer et al., 1982). In addition to extending the available local oscillator operating frequency range, a frequency multiplier also allows one to take advantage of the more favorable characteristics of low-frequency sources: lower noise, larger tuning range, lower cost, and better stability. Excellent results that are near theoretical limits have been reported by Archer, who has built Schottky diode multipliers for radio astronomy receivers. He has demonstrated a tripler that provides more than two milliwatts over the entire band from 200 to 290 GHz (Archer, 1984a) and another that provides in excess of a milliwatt from 260 to 350 GHz (Archer, 1984b).

IV. Optical Scattering System

The design of the multichannel scattering apparatus is based upon previous experience with single-channel systems for laboratory (Park et al., 1980; Yu et al., 1983; Pawley et al., 1985) and tokamak (Semet et al., 1980; Taylor et al., 1980; P. Lee et al., 1982a, 1982b) FIR scattering studies. Although the first multimixer scattering system was built at UCLA for the Microtor tokamak, subsequent design improvements warrant describing in detail the most recent multichannel apparatus constructed (by UCLA) for the TEXT tokamak. Both systems are fundamentally the same with the major physical deviations resulting from the tokamak to which the scattering device must conform.

A. Homodyne Detection Scheme

The FIR scattering apparatus (on TEXT) is shown schematically in Fig. 16. The total height of the system is $\simeq 7$ m and consists of a base, tower,

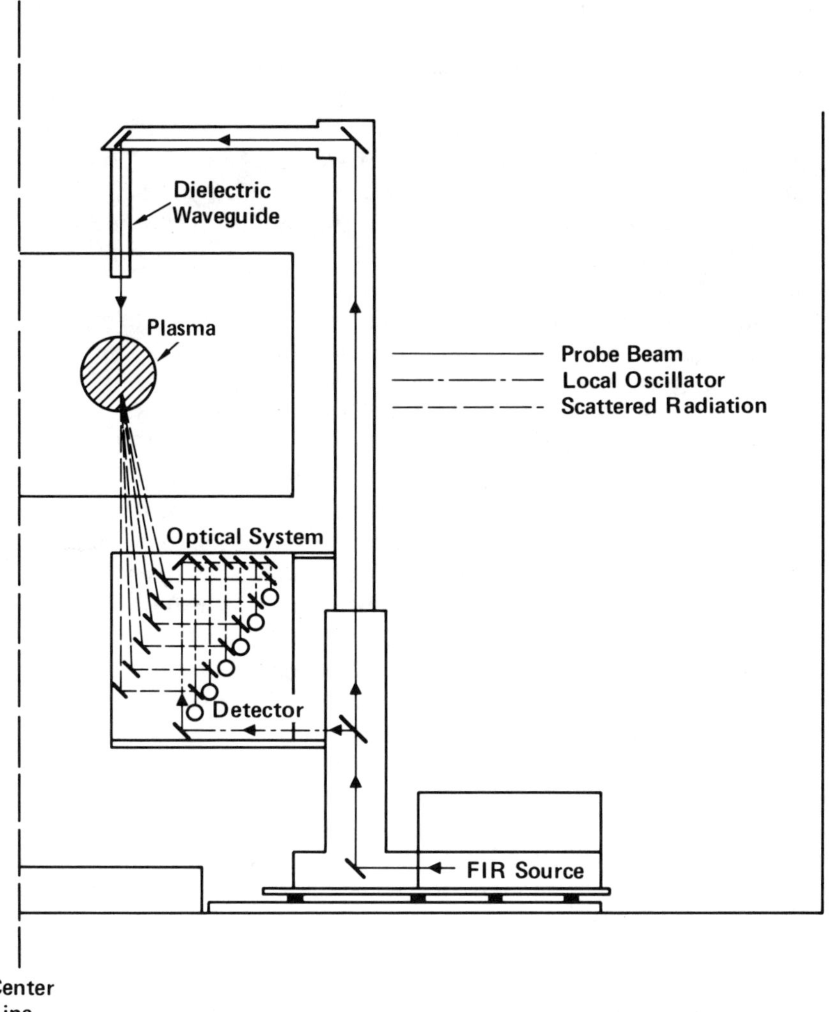

Fig. 16 Schematic of multichannel FIR scattering system on the TEXT tokamak.

and receiver optics table. The base, which serves as a platform for the far-infrared source plus a shielded container for the amplifiers, also provides stability for the tower. A cylindrical stainless steel tower is used to support the beam directing mirrors which bring the probe beam to the top of the tokamak. It also serves as the mount for the receiver optics table (1.2 m × 1.5 m NRC table with component mounts) that holds the six collection mirrors for the scattered radiation along with the optical elements necessary to bring the signal and local oscillator beams to the detectors.

The base rests on precision rails to allow the apparatus (and consequently the scattering volume) to be scanned radially across the plasma cross section. Similar rails are utilized on the cylindrical tower in order to scan the system vertically from plasma top to bottom. These translations permit the entire plasma cross section to be probed except where limited by port access (≃20% of the plasma cross section on TEXT cannot be accessed). Horizontal and vertical movement of the apparatus is achieved via motor drives which are calibrated and remotely controlled from the data acquisition room.

The source beam must propagate ≃ 10 m to reach the plasma, thereby necessitating the use of waveguide techniques in order to maintain a manageable beam size. From detailed studies of waveguides employed in the far-infrared by J. P. Crenn (Crenn, 1979a; Crenn and Veron, 1979), it clearly appears that the circularly oversized dielectric guides are superior to metal guides, mainly because they preserve the linear polarization of the input beam and because their transmission at zero incidence angle is better, provided their diameter is sufficiently large. The beam matching to the waveguide is optimized when the beam waist of the incident radiation is at the input end of the guide, and has a diameter (measured at the e-folding point of the intensity profile) 0.35 to 0.45 times the tube diameter (Crenn, 1979a; Crenn and Veron, 1979). Circular cross section (7.5 cm diameter) dielectric waveguide (plexiglass) is employed for this purpose on TEXT. Laboratory tests have shown that ≃65% transmission of the beam can be expected for the indicated patch lengths with a 1222 μm source. A Gaussian beam with $a_o = 2$ cm (waist, radius at e^{-1} point of the electric field) is matched to the waveguide entrance via use of a polyethylene lens (f.l. = 100 cm) after the output of the FIR laser. By properly matching the input beam to the waveguide, attenuation is minimized (≃35% over the 10 m path) and the exit beam maintains parameters similar to that of the input beam (i.e. $a_o = 2$ cm, linearly polarized). As a result, the attendant difficulty of long propagation path lengths with large machines is solved.

Measurements performed by D. Veron (1980) for interferometer design purposes on the JET (Joint European Torus) tokamak have observed that for a 195 μm beam following a 30 m long waveguide setup with eight right angles, a transmission of 0.55 ± 0.05 could be obtained for a 4 cm I.D. glass tube. Increasing the glass tube size to 6 cm I.D. resulted in an improved transmission of 0.63 ± 0.05. However, for a 7 cm I.D. copper tube, the transmission was only 0.36 ± 0.03. These results are conveniently summarized in Figs. 17 and 18, where at small angles of incidence the glass waveguide possess reasonable transmission characteristics as well as maintaining the polarization state. Referring to Fig. 18, E_M is the magnitude of the electric field in the direction of its largest value and E_m the magnitude in the direction of the smallest value. An additional point to note from Fig. 18 is that it is difficult to maintain linear polarization within an overmoded copper waveguide.

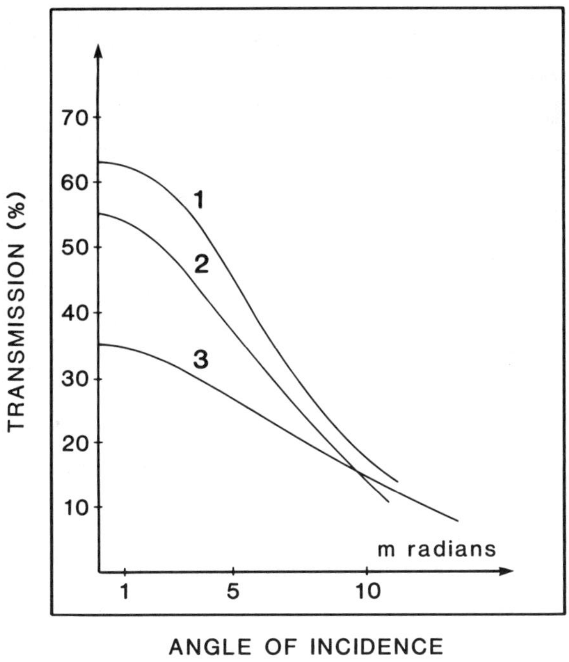

FIG. 17 Transmission of 195 μm radiation through 30 m length of glass and copper waveguide as a function of angle of incidence (D. K. Mansfield, private communication, 1982).

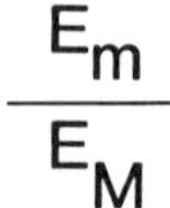

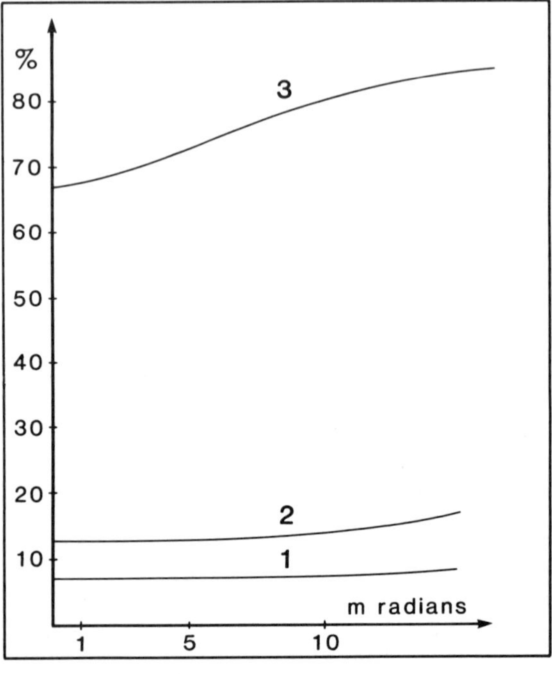

FIG. 18 Variation of polarization as a function of angle of incidence for the configuration described in Fig. 17 (D. K. Mansfield, private communication, 1982).

At certain FIR wavelengths, the long propagation paths also result in unacceptable atmospheric absorption ($\simeq 0.1\ m^{-1}$ absorption coefficient; for example, at 393 μm the loss is $\simeq 1$ dB/m) which requires filling the beam path with dry nitrogen. This is conveniently achieved by the use of some type of waveguide as it serves as a container for the controlled atmosphere (Crenn, 1979a, 1979b; Veron, 1980 and 1982; Marcateli and

Schmeltzer, 1964; Degnan, 1976; Crenn and Veron, 1979 and 1981). Reported total average loss coefficients of $\simeq 2.4 \times 10^{-2}\ \text{m}^{-1}$ have been measured at 195 μm using such a system (Crenn, 1979b).

Now we can proceed to trace out the beam path through the scattering apparatus. From the base of the apparatus the beam is matched to the waveguide and then directed by a mirror, vertically, up the tower. At a distance of approximately 1.5 m from the base a beam splitter is mounted. Here, a portion of the beam is split off via an inductive mesh beam splitter (nickel, 32 lines/cm, reflectivity $\simeq$70%) which constitutes the local oscillator drive for the Schottky diode mixers. The remainder of the beam passes through the beam splitter and continues to the top of the tokamak where it is directed into the plasma by use of two mirrors. A lens is used to locate a new beam waist at the plasma center ($a_0 = 2$ cm). This beam serves as the source or probe beam (also referred to as the incident beam). The incident beam is directed into the tokamak such that $\mathbf{k}_0 \perp \mathbf{B}_\mathrm{T}$ and its polarization, $\mathbf{E}_0 \parallel \mathbf{B}_T$, where $\mathbf{B}_T$ is the toroidal magnetic field.

The local oscillator beam is directed to the receiver optics table where it is divided into six separate beams of nominally equal power via inductive mesh beam splitters. This is illustrated in Fig. 19. The actual meshes (nickel) employed represented reflectivities (lines/cm) of 20% (8), 30% (12), 40% (16), 50% (20), and 100% (mirror), in that order with respect to the incoming local oscillator (LO) beam. Since each mixer did not receive an equal portion of the LO power, a calibration technique, to be described in Section V, accounted for any discrepancies.

Also located on the receiver optics table are six collection mirrors which direct the scattered radiation to the six, respective, detectors. The orientation of the collection mirrors in Fig. 19 is such that the scattered radiation is directed away from the torus center (SAT). The mirrors may also be configured so that the scattered radiation will be directed towards the tokamak (STT); i.e. reflected about the 0° scattering channel. Both arrangements are utilized in these measurements in order to make maximum use of the available port space.

A 50% reflecting beam splitter (nickel mesh, 20 lines/cm) then couples the scattered radiation and local oscillator beams into the detector, focused by a high density polyethylene lens ($F = 0.5$, 9 cm diameter). These two beams mix in the detector resulting in an IF signal at the difference frequency. This IF frequency corresponds to the frequency of the plasma density fluctuations. Such a scheme does not produce any information concerning the direction of propagation of the fluctuations and is referred to as homodyne detection. In Section IV.B, a heterodyne detection technique will be described which permits the wave propagation direction to be ascertained.

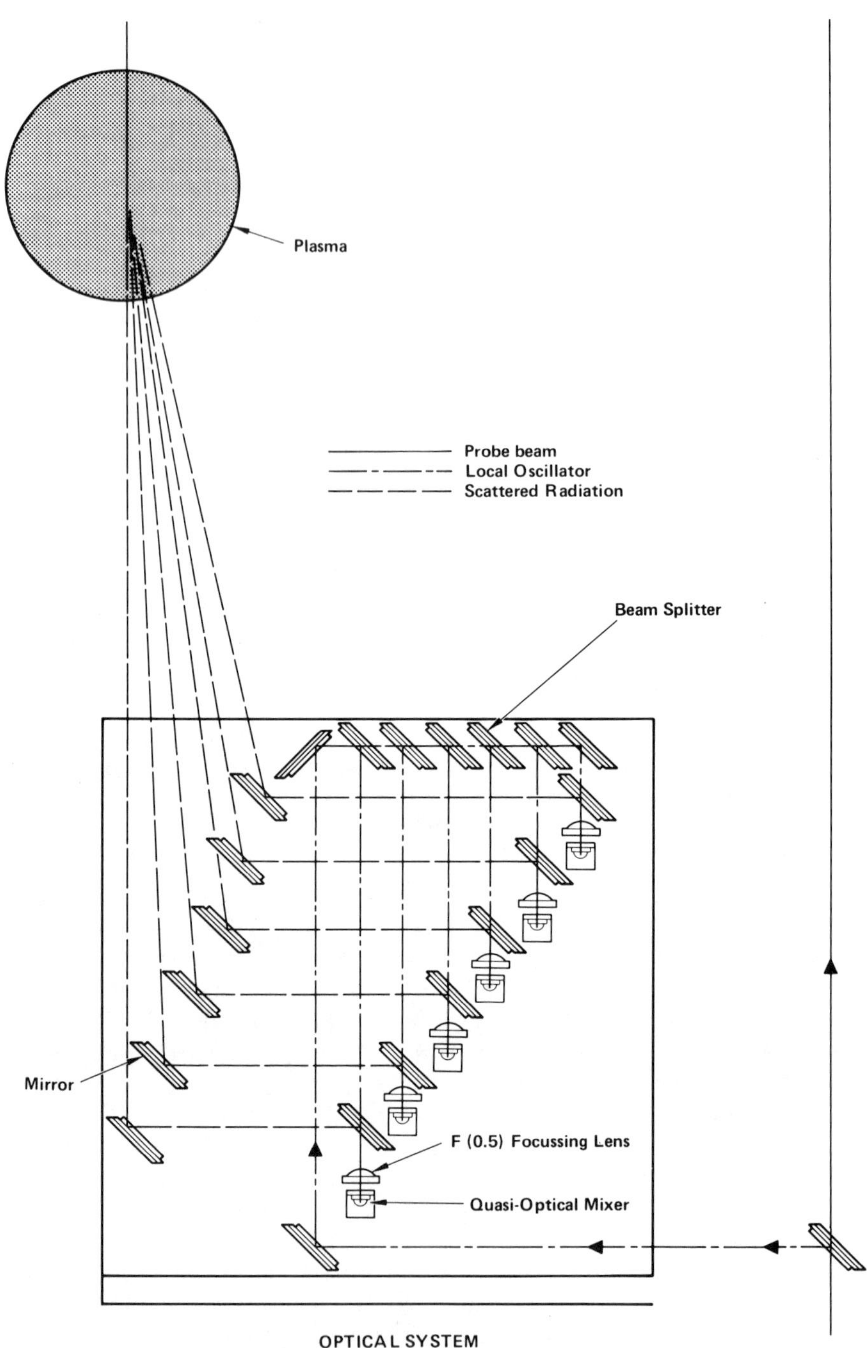

FIG. 19 Multichannel FIR scattering system homodyne receiver schematic.

The collection mirrors for the scattered radiation are positioned to resolve the scattering angles; $\theta_s \simeq 0°$, 2.3°, 4.8°, 7.8°, 10.2°, and 13.1° which for a source at 1222 μm (496 μm) corresponds to wavenumbers of 0, 2.1(4.6), 4.3(10.6), 7.0(17.2), 9.2(22.5), and 11.7(28.9) cm^{-1}, respectively. The 0° channel represents far-forward scattering (Evans et al., 1982) within the divergence of the incident beam and can also be used for interferometric measurements. The wavenumber resolution, Δk, is independent of $\mathbf{k}$ and is $\simeq 2/a_0$ or roughly $\pm 1\ \text{cm}^{-1}$. The waist of the incident (object) beam at plasma center, a_0, is also the spatial resolution perpendicular to the direction of propagation. Spatial resolution along the direction of propagation, determined by the scattered (image) beam, is inversely proportional to the scattering angle, $L_v \simeq 2a_0/\sin\theta_s$ where L_v is the length of the scattering volume. The image beam has waist a_0 at the plasma center.

The probe and scattered beams are plotted in Fig. 20, for the six wavenumbers of interest assuming a 1222 μm source. Both beams are characterized by a Gaussian profile. Their intersection gives an indication of the length of the scattering volume at the e^{-1} falloff point of the scattered power.

Spatial resolution along the direction of the incident beam is essentially a chord average for $k \leq 4.3\ \text{cm}^{-1}$ ($L_v = 45$ cm). The three channels with the largest wavenumbers, $k = 7$, 9.2, and 11.7 cm^{-1} possess scattering volume lengths of $\simeq 27$, 23, and 16 cm, respectively, which are to be compared to the TEXT minor diameter of 54 cm. Experimental verification of L_v is obtained by use of an acoustic cell (see Section VI.) as well as by actual tokamak measurements (see Section VII.).

B. Heterodyne Detection Scheme

The detection scheme outlined in the previous section was for homodyne scattering: a technique which does not provide information concerning wave propagation direction. In this section, a heterodyne detection scheme will be described which makes possible the determination of wave propagation direction, $\mathbf{k}$, for low-frequency (i.e. ≤ 1 MHz) fluctuations. The resulting changes to the experimental apparatus, from the homodyne setup, are shown in Fig. 21.

After the FIR laser, a beam splitter is required to divide the source into two legs. One leg passes through a half wave plate, is focused by a polyethylene lens (f.l. = 100 cm) and reflected via a polarized beam splitter (stainless steel, 158 lines/cm grid) into the dielectric waveguide. The other leg is passed through a lens and focussed onto a rotating grating which frequency shifts the beam by an amount $\Delta\omega_g$. This phase modula-

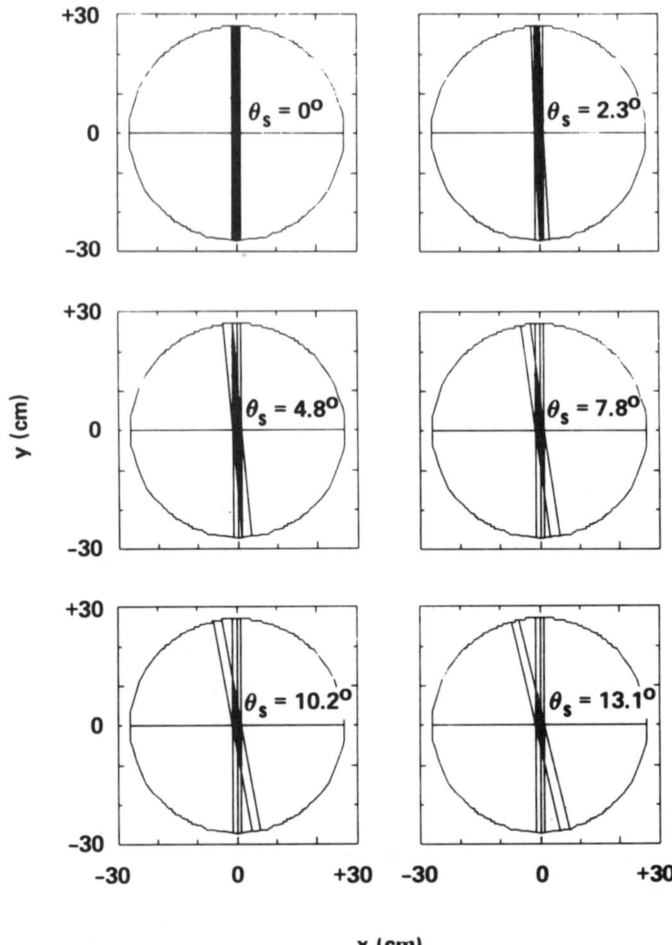

FIG. 20 Scattering volume length versus scattering angle (θ_s) for a source beam at λ_o = 1222 μm.

tion technique is thoroughly described by Veron (1979). The frequency shifted beam is then recollimated, focused, passed through the polarized beam splitter, and introduced with the oppositely polarized leg into the waveguide. At this point we have two perpendicularly polarized beams at a difference frequency, $\Delta\omega_g$.

The beam splitter located in the tower must now be replaced with a polarizing beam splitter (stainless steel, 158 lines/cm) which selectively reflects the frequency shifted beam for use as the local oscillator. The

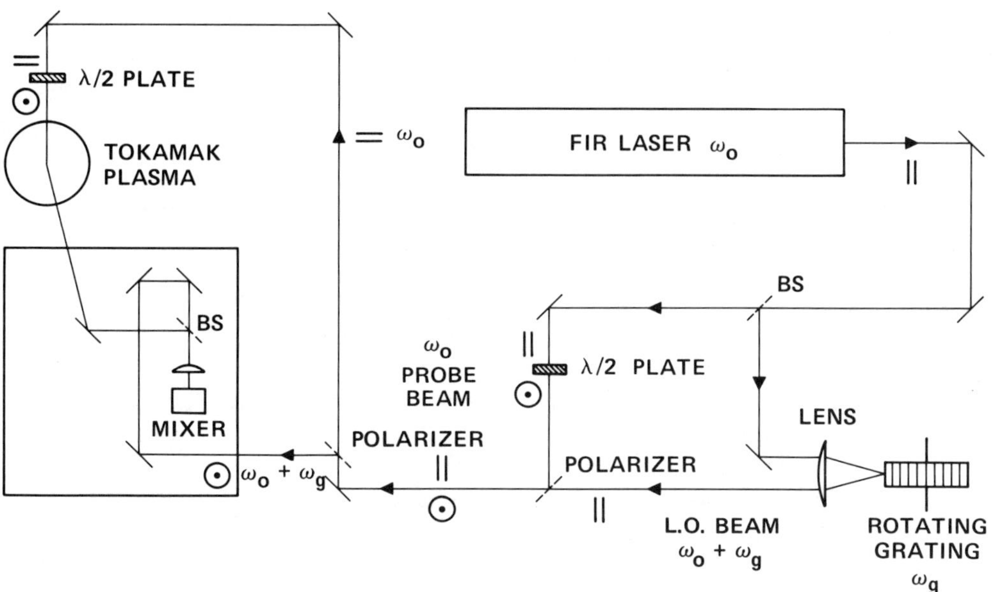

FIG. 21 Schematic of heterodyne receiver system employing a rotating grating; (a) rotating grating arrangement, and (b) receiver optics arrangement.

remaining beam propagates through the tower, passes through another half wave plate (thus returning it to its original polarization) and is sent into the plasma to serve as the probe beam. Now when these two beams combine in the mixer the difference frequency is no longer that of the plasma fluctuation, $|\pm\omega|$. Rather, it is equivalent to $|\Delta\omega_g \pm \omega|$ where ω is the frequency of the density fluctuation. Now if $\Delta\omega_g > \omega$ (ideally $\Delta\omega_g \gg \omega$), the direction of propagation of the fluctuation can be determined, i.e. it is observed as either an up shift or down shift, depending on **k**.

For the rotating grating technique, one may typically operate with a frequency shift of $\Delta\omega/2\pi \simeq 85$ kHz for source radiation at 1222 μm ($\Delta\omega \propto \lambda_o^{-1}$). It is possible to double $\Delta\omega/2\pi$ ($\simeq$170 kHz) with conventional motors for a short period of time, however, under these conditions the motor tends to overheat and air turbulence from the high speed rotation becomes significant. When the rotating grating scheme is implemented into the system, due to inefficiencies in reflection from the grating and the additional optical elements, only a two-channel system is realized (this was only done at 1222 μm). Depending upon the optical quality of the grating, reflection efficiencies of $\simeq$80% have been observed for linearly polarized radiation with its electric field in the plane of incidence (Veron, 1974).

An additional modulation technique which has been successfully employed in an interferometer application is to reflect a portion of the laser beam from a linearly translating corner mirror as shown in Fig. 22 (Baker et al., 1978; Baker, 1980), for the case of the Doublet III tokamak system. In such a system, the velocity of the corner reflector may be $\simeq$6 m/s which corresponds to $\Delta\omega/2\pi \simeq 10$ kHz ($\Delta\omega \propto \lambda_o^{-1}$). As will be seen in Sec. VII.A., microturbulence induced plasma fluctuations are observed at frequencies up to $\omega/2\pi \simeq 1$ MHz for large wave vectors. Hence, $\Delta\omega_g \not> \omega_{\text{fluct}}$ making both the rotating grating or translating mirror systems marginal for heterodyne detection measurements. Due to the low cost of a rotating grating system, it was employed for the initial measurements on TEXT. However, improved techniques which are currently being applied at TEXT will be described in the ensuing paragraphs. Nevertheless, the rotating grating scheme is an ideal modulation technique for interferometry and hence another reason for its use on TEXT.

In the case of optically pumped lasers, an alternative method exists for obtaining the appropriately frequency shifted beam. Here one employs two FIR lasers whose cavity lengths are adjusted to provide a frequency offset within the frequency range of the FIR lasing transition (Wolfe et al., 1976; Mansfield et al., 1980; Ma et al., 1982; Yamanaka et al., 1980). A typical arrangement is illustrated in Fig. 23, where the output of a single CO_2 laser pumps the two FIR lasers employed in the JIPP T-II tokamak interferometer (Yamanaka et al., 1980). Such systems can usually provide frequency offsets of order of several megahertz. Although only a single CO_2 pump laser was utilized in the system shown in Fig. 23, one may also employ two separate pump lasers to increase the FIR output power level (Ma et al., 1982). However, one must then ensure that the CO_2 pump lasers exhibit the proper stability. (This point turns out to be a very critical one.)

The above mentioned techniques provide a simple and convenient method for obtaining an IF frequency in the megahertz region. However, considerable effort is required to properly lock the CO_2 and FIR lasers so that the beat mode is stable and narrowline (Cheo, 1985). Therefore, there continues to be interest in other techniques for providing such stability. The need for a stable IF in the 1 MHz range has resulted in the newly developed "double barrel" FIR laser (Lehecka et al., 1985). Although rotating gratings, translating mirrors, and separate FIR lasers can be employed to generate such an IF, they suffer from efficiency and stability limitations. Both of these problems are eliminated by the 'double barrel' laser (see Fig. 24) which employs two FIR lasers within the same cavity structure thereby compensating for temperature-induced cavity length changes, etc. Thus far, an IF stability of $\pm$10 kHz has been achieved in a

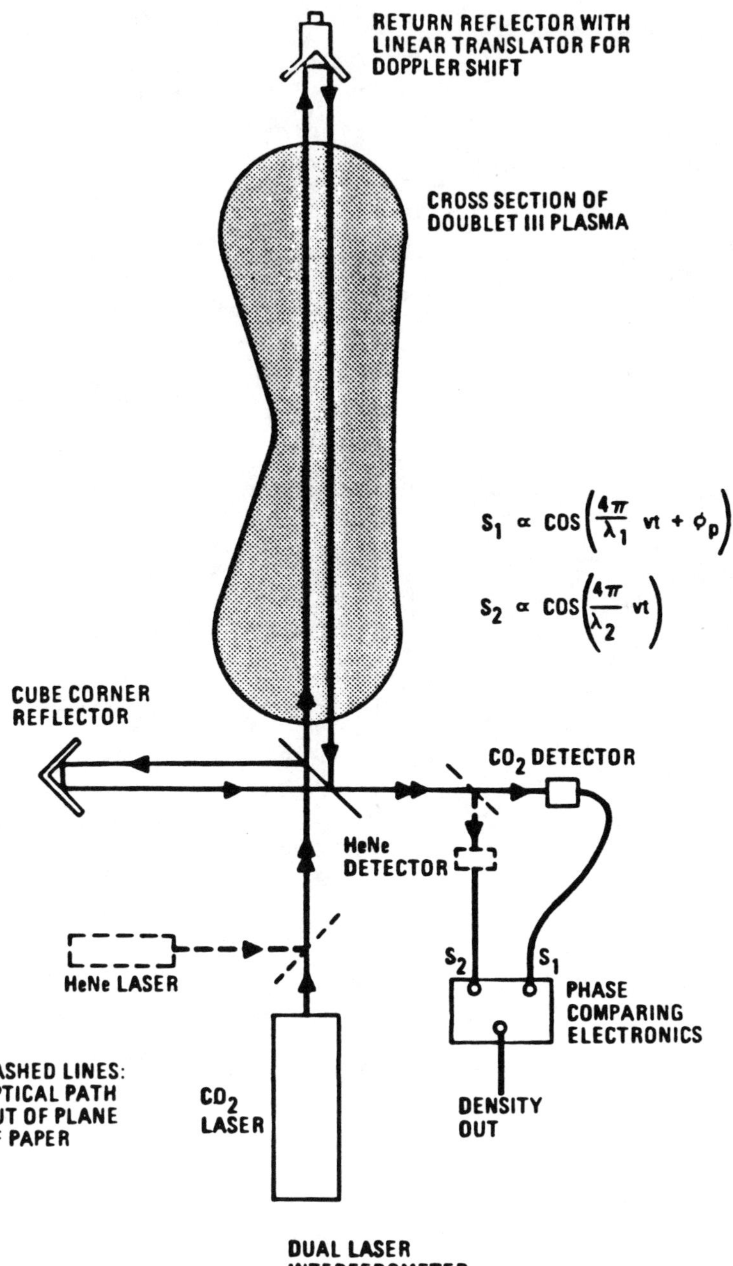

FIG. 22 Schematic of the GA Technologies Doublet III CO_2 laser interferometer showing a moving corner reflector (Baker and Lee, 1978).

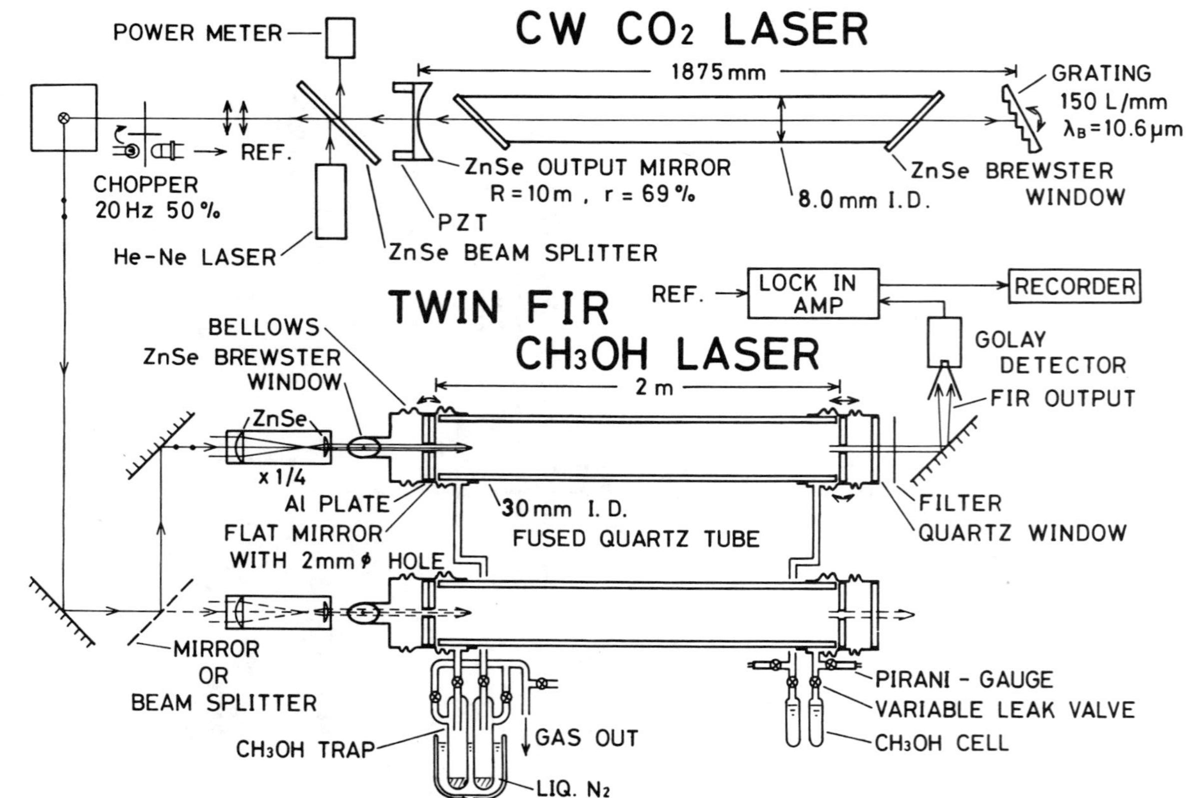

FIG. 23 Schematic of a CO_2 optically pumped twin CH_3OH FIR laser operating at 199 μm with greater than 1 MHz modulation frequency (Yamanaka et al., 1980).

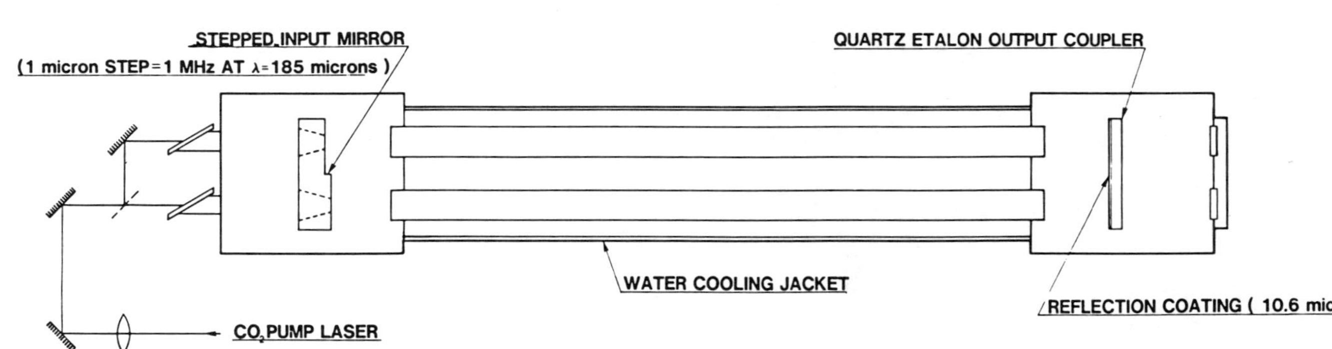

FIG. 24 Schematic of single cavity dual frequency FIR laser dubbed the "double barrel" laser.

free-running state. In addition, the spatial mode quality efficiently transforms into the Gaussian free space mode. Continuing research is ongoing at UCLA to maximize the power output from such an arrangement and to look at the viability of other laser configurations.

Acousto-optic modulators provide another means whereby the frequency of a laser beam may be shifted. To date, commercial standards (60–80% efficient) have only been achieved for wavelengths up to $\simeq 10\ \mu m$ (Kristal, 1978; Hugenholtz and Meddens, 1979; Jacobsen and Jolin, 1981; Jacobsen, 1978; Hugenholtz and Meddens, 1982). In the far-infrared, the efficiencies at present are much lower with the highest reported being $\simeq 1.5\%$ (Vogel and Dodel, 1984). Such a system consists of an acoustic cell (some type of medium which has good transmission in the FIR, either solid or liquid; see Section VI) upon whose surface a piezo-electric transducer (PZT) is attached. These are typically operated in the Bragg regime where all of the diffracted light appears along the first order. Acousto-optic modulators can produce offset frequencies in the megahertz range. If the efficiency can be improved, such a system would be a low-cost alternative to the two FIR laser configuration while providing the frequency difference desired but not attainable with the rotating grating technique.

Finally, a new method for measuring the wave propagation direction which is both economical and reliable has been developed by Tsukishima et al. (1978, 1980, 1984) and Asada et al. (1980, 1981). This technique, referred to as "homodyne spectroscopy", employs two homodyne scattering signals which are phase shifted relative to one another by $\varphi = \pi/2$. Two such signals may be processed using digital time series analysis techniques such as the auto and cross power spectra in order to produce information on the blue–red sidebands of the scattered radiation. The IF output signal from the mixer in the homodyne configuration produces a signal which is of the form

$$V(t) = \mathrm{Re} \int_{-\infty}^{+\infty} \frac{d\omega}{2\pi} N_{+}(\omega) \exp\left[i(\omega t + \varphi)\right] \tag{15}$$

where $N_{+}(\omega) = gE_1E_o n(k_{+}, \omega)$ with g being a constant. We can define two independent IF signals $V_1(t)$ and $V_2(t)$ by putting $\varphi = 0$ and $\varphi = \pi/2$ in Eq. (15), respectively. Then, introducing an analytical signal $v(t)$, Eq. (15) may be written as

$$V(t) = V_1(t) - iV_2(t) = \int_{-\infty}^{+\infty} \frac{d\omega}{2\pi} N_{+}(\omega) \exp\left[i\omega t\right]. \tag{16}$$

The Fourier transform of Eq. (16) yields $N_{+}(\omega)$ in terms of the measured quantities $V_1(t)$ and $V_2(t)$. The power spectral density is then obtained from

$$S_{\pm}(\omega) = \frac{|N_{\pm}(\omega)|^2}{2T} = G_{11}(\omega) + G_{22}(\omega) \pm [G_{12}(\omega) - G_{21}(\omega)], \qquad (17)$$

where T is the record length of $V_1(t)$ and $V_2(t)$. $G_{ii}(\omega)$ and $G_{ik}(\omega)$ with $i \neq k$ are the estimates of the auto and cross power spectral densities of the sample records of $V_1(t)$ and $V_2(t)$,

$$G_{ii}(\omega) = 4\int_0^\infty \left(1 - \frac{|\tau|}{T}\right) R_{ii}(\tau) \cos \omega\tau \, d\tau, \qquad i = 1,2, \qquad (17a)$$

$$G_{ik}(\omega) = 4\int_0^\infty \left(1 - \frac{|\tau|}{T}\right) R_{ik}(\tau) \sin \omega\tau \, d\tau, \qquad i \neq k, \qquad (17b)$$

with

$$R_{ik}(\tau) = \frac{1}{T - |\tau|}\int_0^{T-|\tau|} V_i(t + \tau)V_k(t)\, dt, \qquad i,k = 1,2. \qquad (17c)$$

The parameters $R_{ii}(\tau)$ and $R_{ik}(\tau)$ are the estimates of the auto and cross correlation functions of $V_1(t)$ and $V_2(t)$, respectively. For the details of the calculation of these parameters the reader should refer to Asada et al. (1980). The parameters $G_{ii}(\omega)$ and $G_{ik}(\omega)$ may be calculated by employing the well developed digital time series analysis techniques (Smith et al., 1974).

The homodyne spectroscpoy technique (Tsukishima et al., 1984) has been successfully tested with simulated electrical circuits (Asada et al., 1980), microwave scattering from electrostatic ion waves (Tsukishima et al., 1980; Asada et al., 1981), and with an FIR laser system using two frequency-shifted laser beams (Tsukishima et al., 1984). The first known test of this technique on tokamak microturbulence measurements are being carried out on the TEXT tokamak (Brower et al., 1986). A schematic of the experimental arrangement is shown in Fig. 25. The two-channel homodyne system employs a piece of polyethylene as the phase shifter to produce the $\varphi = \pi/2$ between $V_1(t)$ and $V_2(t)$. The phase shifter is mounted on a rotation stage which facilitates easy tuning of φ. The preliminary results are positive and will be shown in Section VII. Such a system provides an easily implemented and economical alternative to the previously discussed heterodyne schemes.

In addition to scattering, the 0° channel may be used for far-forward scattering (Evans et al., 1982) or for interferometry. Both possibilities are utilized. For interferometric measurements one could use homodyne techniques and manually count the density fringes or employ a heterodyne setup which, when compared with a reference signal, provides a direct measurement of the phase or plasma density. The reference signal consists of mixing the shifted and unshifted components of the source beam to obtain a signal at the difference frequency, $\Delta\omega$. These measure-

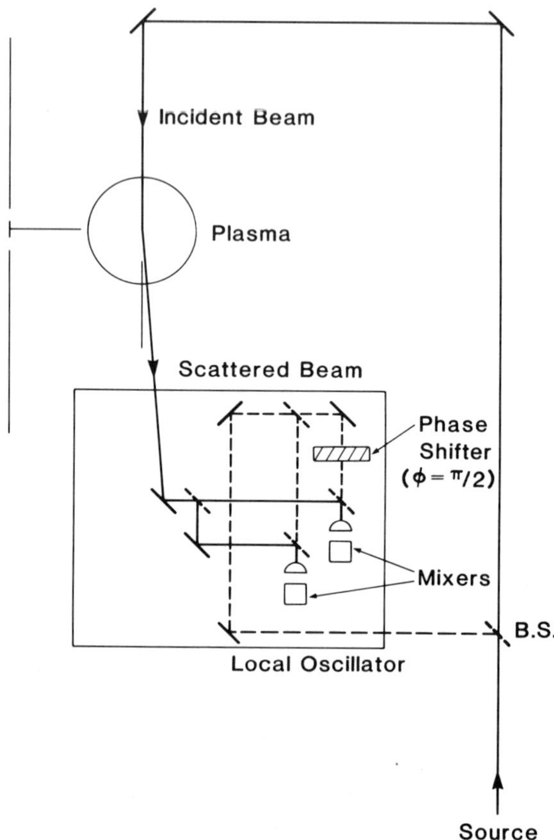

FIG. 25 Schematic of optical arrangement for homodyne spectroscopy measurements.

ments are performed and will be shown in Section VIII. The technique used to determine the phase between the probe and reference beams is a digital one referred to as complex phase demodulation (Joyce et al., 1985). Digital complex demodulation is a digital version of analog heterodyne demodulation and allows the simultaneous measurement of the amplitude and phase modulations of a narrowband modulated carrier signal as a function of time.

C. Multispatial Configuration

In addition to the possibility of multiangle scattering measurements, the system may be configured to provide a multispatial scattering capability. Such a system is extremely useful for studies such as the recent mode converted ion Bernstein wave measurements on the UCLA Microtor tokamak (A. Y. Lee et al., 1982; Park et al., 1984a). The configuration, illustrated schematically in Figs. 26(a) and (b) for both horizontal and vertical distributions, provides the capability of simultaneously observing the time history of the waves at three different spatial locations. In Fig. 26(a), beam splitters are used to produce probe beams at three different spatial locations. The scattered beams are collected by rotatable collection mirrors mounted on translation stages. In Fig. 26(b), the three collection mirrors are situated such that they are looking vertically along a single probe beam thereby providing the top-midplane-bottom measurement locations.

D. Far-Forward Scattering

In the preceding sections, it was tacitly assumed that the scattering angle was larger than the divergence angle of the input probe beam. The opposite limit is also of interest for plasma diagnostics and has been referred to as far-forward scattering, ultra-forward scattering or as homodyne scattering (Slusher and Surko, 1980; Evans et al., 1982; Doyle et al., 1983; Sonoda et al., 1982; Yu et al., 1983). A related technique is the use of "phase scintillations" of a probing electromagnetic wave as in radio astronomy (Lovelace et al., 1970).

Far-forward scattering depends on the fact that an electromagnetic wave traversing a refractive medium such as a plasma undergoes phase and amplitude modifications which are related to the wavelength, position, amplitude, and frequency of phase fluctuations in the medium. The simplest case to treat is that of an incident coherent Gaussian beam (Evans et al., 1982). The geometry is illustrated in Fig. 27. The laser beam propagating in the z direction is diffracted by the plasma wave (propagating transverse to the beam in the x direction) which acts as a moving sinusoidal phase grating $\varphi(x) = \Delta\varphi \sin z(kx - \omega t)$. As a result of passage through the wave, the beam acquires an intensity component at the wave frequency and successively smaller components at harmonics of the wave frequency. The component at the wave frequency, i.e., the first-order diffracted beam, corresponds to the IF signal from the heterodyne detector. It can be shown (Yu et al., 1983) that the amplitude I_1 and phase φ_1 of this component are given by

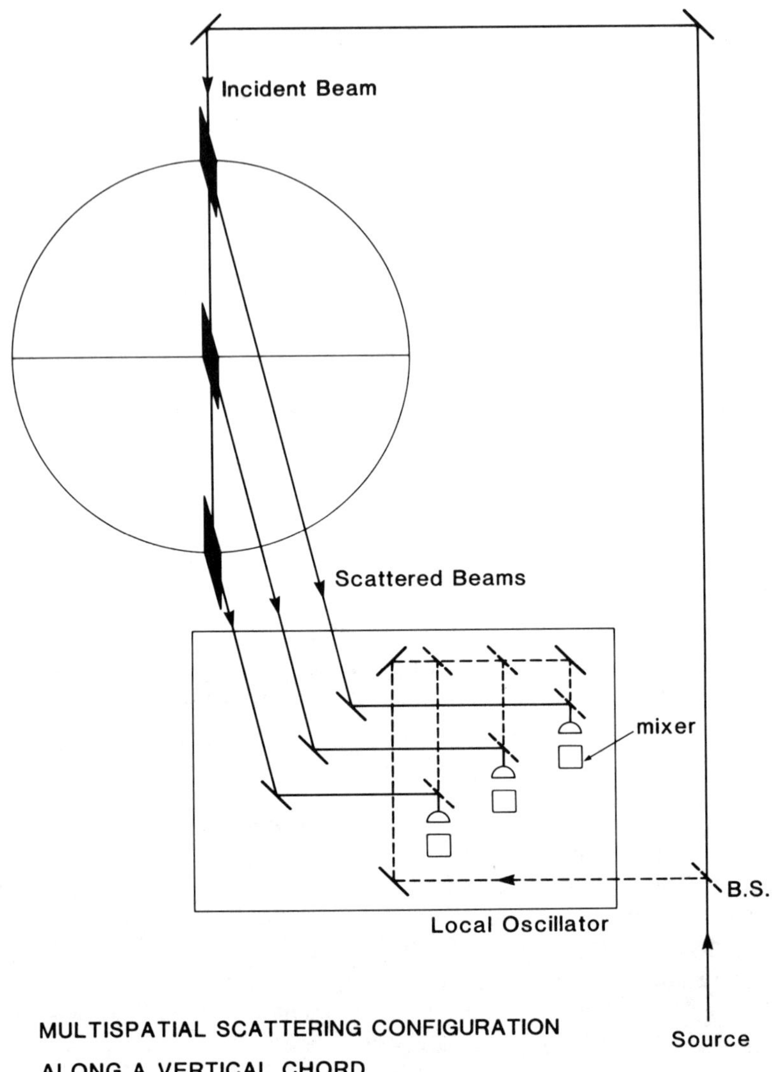

FIG. 26 Experimental arrangement for the FIR multispacial scattering system; (a) horizontal (Park et al., 1982), and (b) vertical.

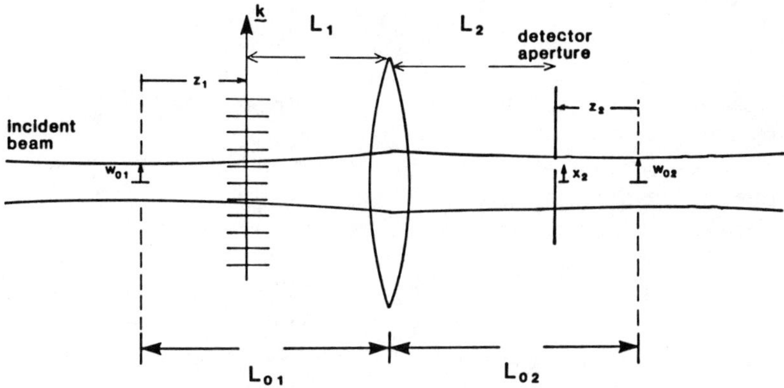

FIG. 27 Schematic of arrangement for far-forward scattering (Luhmann and Peebles, 1984).

$$I_1 = \frac{P_0\Delta\varphi}{\pi\omega_2^2} e^{-(u_x^2+u_y^2)} e^{-(v^2/2)} (e^{2u}x^v + e^{-2u}x^v - 2\cos\rho v^2)^{1/2}, \quad (18)$$

$$I_1 \sin\varphi_1 = \frac{P_0\Delta\varphi}{\pi\omega_2^2} e^{-(u_x^2+u_y^2)} e^{-(v^2/2)}$$

$$\times \left[e^u x^v \cos\rho v\left(u_x - \frac{v}{2}\right) - e^{-u}x^v \cos\rho v\left(u_x + \frac{v}{2}\right)\right], \quad (19)$$

where P_0 is the power of the incident beam.

$$u_x = \frac{x_2}{\omega_2}, \quad u_2 = \frac{y_2}{\omega_2}, \quad v = \frac{\omega_1 k}{1 + \rho^2},$$

and

$$\rho = \frac{z_1}{z_{r1}} + k_o\omega_1^2 \frac{L_1 - L}{L_1^2}.$$

The parameters ω_1 and ω_2, the $1/e$ radii of the laser beam intensity profiles at the waves and the detector, respectively, are related to the beam waist radius ω_{01} by

$$\omega_2 = \left(\frac{L_1L_2}{L}\right)\frac{(1+\rho^2)^{1/2}}{\omega_1 k_o}, \quad \omega_1 = \omega_{01}\left[1 + \left(\frac{z_1^2}{z_{ri}^2}\right)\right]^{1/2},$$

where $z_{ri} = k_o\omega_{01}^2$ is the Rayleigh length and

$$L = \left(\frac{1}{L_1} + \frac{1}{L_2} - \frac{1}{f}\right)^{-1},$$

where L_1 and L_2 are the distances from the lens of focal length f to the waves and the detector, respectively, as shown in Fig. 27. One may, therefore, determine the wavenumber of the plasma waves by measuring the position of the peak scattered signal and the ratio of the scattered signal on axis to the peak signal as illustrated in Fig. 28. Doyle et al. (1983) have studied low-frequency fluctuations in the TOSCA tokamak using such far-forward scattering of a CO_2 beam. However, as shown by Yu et al. (1983) in a far-infrared laboratory experiment, even minute departures from the Gaussian beam assumption can lead to problems in interpreting the scattering data.

V. Mixers, Amplifiers and Data Acquisition

The detection schemes previously described (i.e. homodyne and heterodyne) both involve signal and local oscillator beams mixing in a receiver, resulting in an output signal at the difference frequency. This signal is then amplified, digitized, and analyzed on a computer system. The entire system is schematically outlined in Fig. 29. The signal and local oscillator beams are at a high frequency (RF). The mixing element (GaAs Schottky barrier diode) is biased to the nonlinear range of its diode characteristic curve by a small dc bias (0.2 mA) and the local oscillator (LO

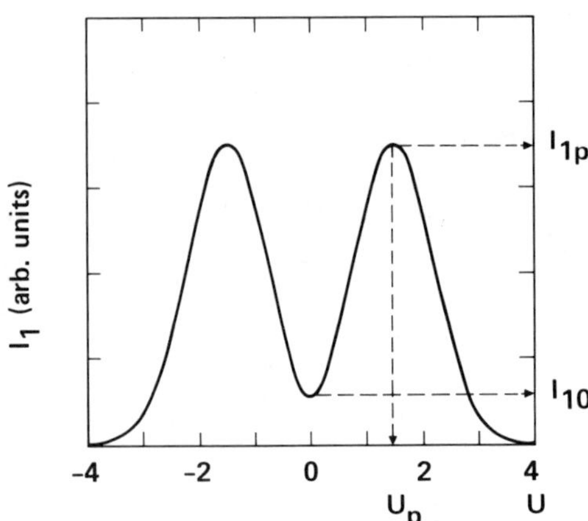

FIG. 28 Theoretical dependence of the scattered intensity I_1 on the normalized position U of an apertured detector which is scanned across a laser beam which has passed through the region of wave activity (Luhmann and Peebles, 1984).

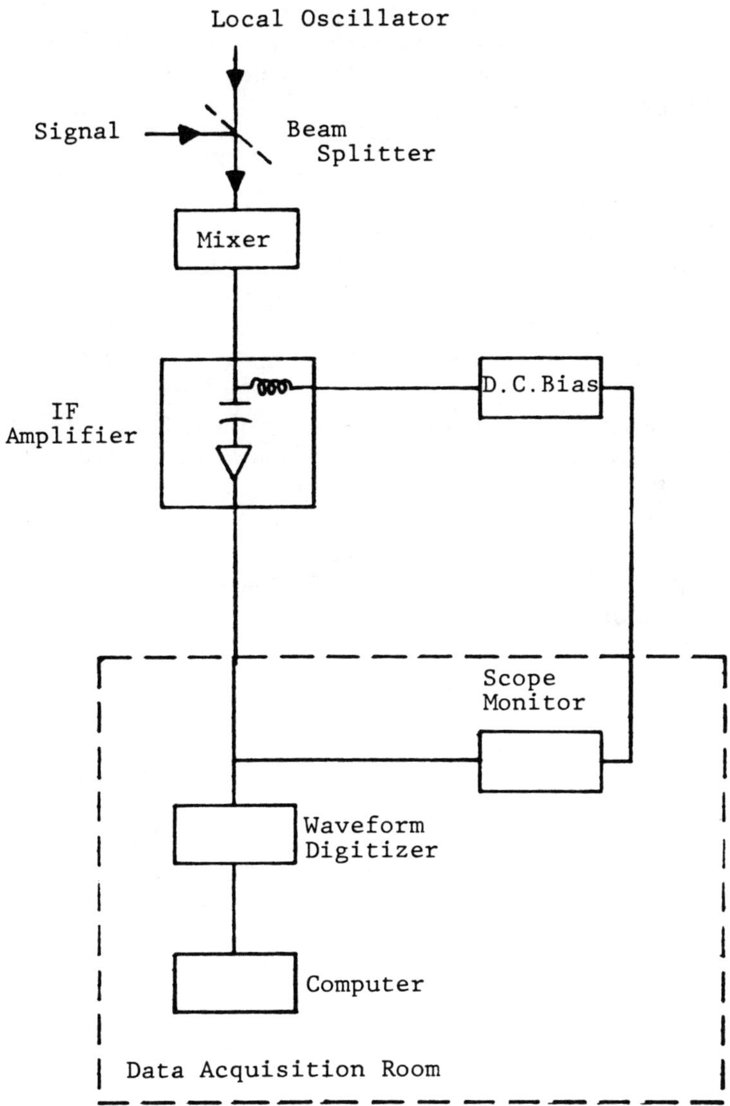

FIG. 29 Schematic for signal processing system.

bias $\simeq 0.2$ mA). The RF signal beam, several orders of magnitude lower in power than the local oscillator, is combined with the LO in the mixer which provides the low-frequency output signal (IF).

The heart of the above-mentioned receiver system is a low-noise mixer. For microwave systems, conventional fundamental mode waveguide mixers are utilized. Unfortunately, at the highest frequencies of interest, mechanical tolerances impose a severe impediment to fundamental waveguide mixer fabrication. Although low-noise waveguide mixers have been demonstrated at frequencies as high as 600 GHz (Erickson, and Fetterman, 1982), the cost prohibits their use in a multichannel Scattering system. Hence, there has been considerable interest in so-called quasi-optical structures which serve to eliminate some of the fabrication problems as well as possessing low losses in the submillimeter region.

The first quasi-optical mixer configuration to demonstrate low-noise performance in the far-infrared region was the biconical amount. This design was originally conceived by J. J. Gustincic (1977a, 1979b) for a NASA-JPL atmospheric radiometry program and later extended to the FIR for radio astronomy and plasma diagnostic applications (Gustincic et al., 1977). The Schottky barrier diode mixer is quasi-optical with a bioconical antenna design as shown in Fig. 30. Incident radiation is coupled into the mixer via a $3\lambda_0$ aperture in the backshort which then serves to focus the radiation on the whisker (antenna). Polarization components parallel to the receiving antenna create an electric field and the resulting current is in turn rectified at the junction. The chip and whisker are located at the apex of two terminal biconical transmission lines separated by $3\lambda_0$. Choke grooves are placed in the cone bodies to prevent RF leakage. An adjustable backshort section ($3\lambda_0$ diameter) permits tuning of the resonant cavity.

An alternative approach is based on the familiar traveling long wire antenna. To produce a focussed beam from this configuration Krautle et al. (1977, 1978) added a corner reflector. Figure 31 shows a realization of such a mixer by Fetterman et al. (1978). The structure is easily analyzed by utilyzing standard imaging techniques and can be thought of as a simple four-element phased array. This design has proven to work well and has demonstrated low-noise performance at frequencies in excess of 1 THz. Table 2 shows representative corner cube mixer performance obtained by Fetterman and his colleagues at the MIT Lincoln Laboratory (Fetterman et al., 1978, 1980). As can be seen, even at frequencies as high as 2.7 THz, adequate mixers exist for scattering measurements.

Here it should be noted that near-millimeter wave mixer technology is rapidly advancing and that discrete whisker contacted mixers will be replaced by planar mixer arrays within a few years. Specifically, both

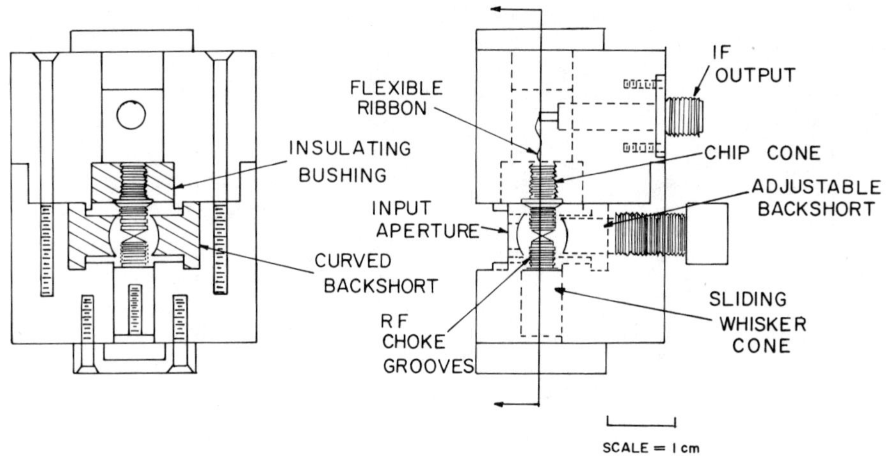

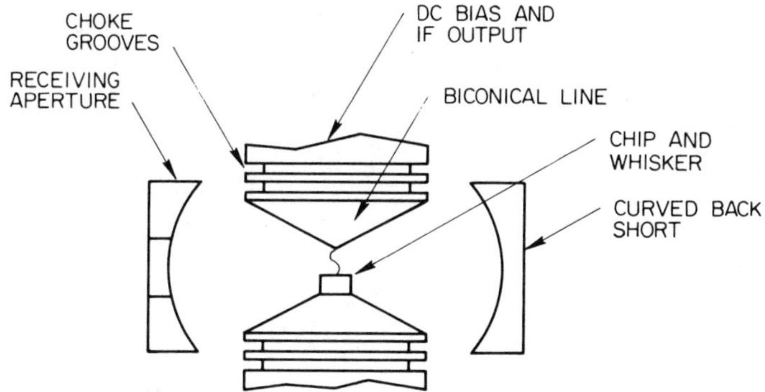

FIG. 30 Schematic of Schottky barrier diode mixer (Gustincic, 1977a).

TABLE II

NOISE TEMPERATURE MEASUREMENTS

λ (μm)	T_s (Room temperature)(K) (Vacuum box, TPX window)	T_s (cooled to 41 K)
432.6	5700	3800 K Temperature mixer = 2900 K T_D = 262 K (with L.O.) Temperature if with isolator 156 K Conversion loss = 10.5 (dB)
432.6	4200	
184	19000	
119	32000	

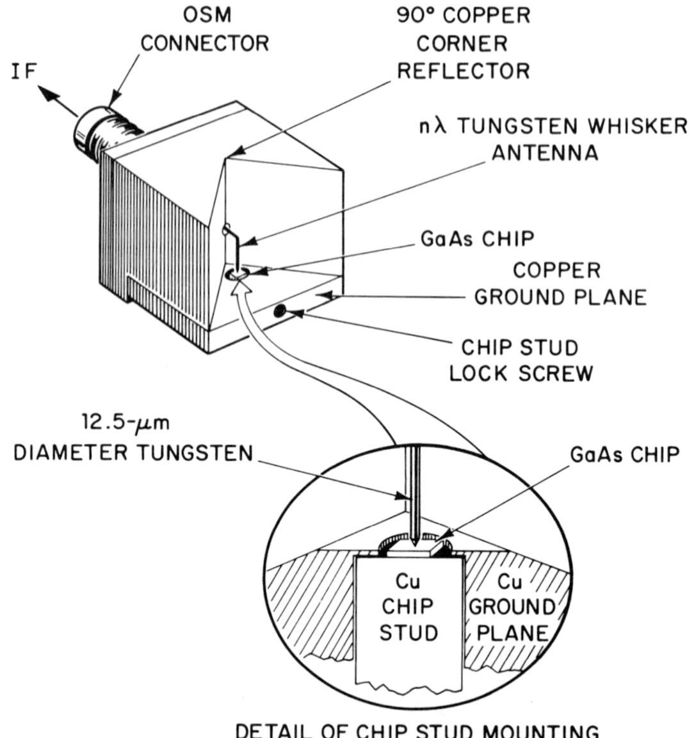

FIG. 31 Schematic of quasi-optical corner reflector Schottky diode mixer (Fetterman et al., 1978).

hybrid and monolithic mixer technology have undergone dramatic improvements during the last few years. Figures 32 and 33 show photographs of a nine-element monolithic diode array fabricated at the California Institute of Technology (Zah et al., 1984) which has established heterodyne system noise temperature of ≃5000°K at 94 GHz. The performance is currently limited by the diode capacitance, and the authors feel that comparable performance will be possible at least up to 300 GHz. Such an array will be particularly useful for the far-forward scattering techniques. An alternative approach has been employed by Fettermen and Chew, 1985 and Parrish et al., 1982, as shown in Fig. 34. Here a high frequency beam lead diode has been bonded onto a planar twin dipole antenna.

FIG. 32 Photographs of monolithic Schottky diode imaging array.

In the actual experimental arrangement for plasma fluctuation measurements, the mixer is mounted (isolated) in an aluminum box, in addition to the diode-whisker brass housing, thereby serving as a shield from external noise radiation. Such extreme shielding measures are required in the harsh tokamak environment, where significant external noise sources such as ICRH, ECRH, synchrotron radiation, and neutral beam heating systems may exist. The only external opening is the $3\lambda_0$ aperture utilized to couple in the signal and LO radiation. Since the receiver system directly views the plasma, noise contributions the plasma frequency [$\omega_p = (4\pi n_e e^2/m_e)^{1/2}$] and the electron cyclotron frequency ($\omega_{ce} = eB/m_e$) are also coupled into the detector and may serve as a substantial background source. This aperture may be shielded by employing a dichroic plate (Chen, 1973; P. D. Potter) which has been used in the past on NASA-JPL

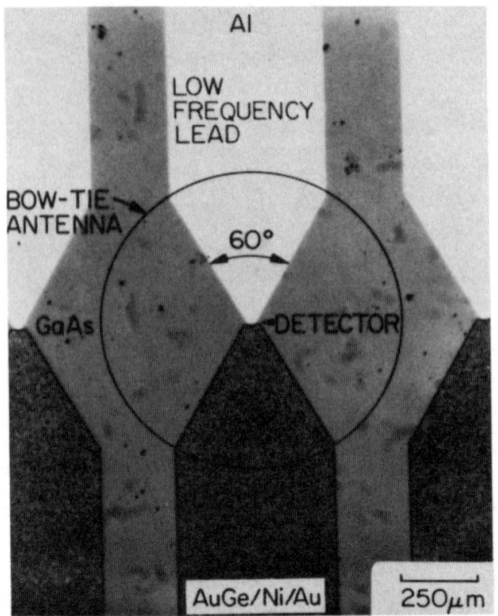

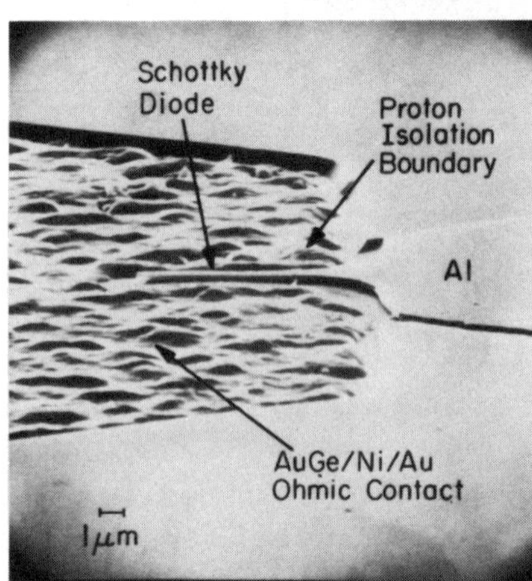

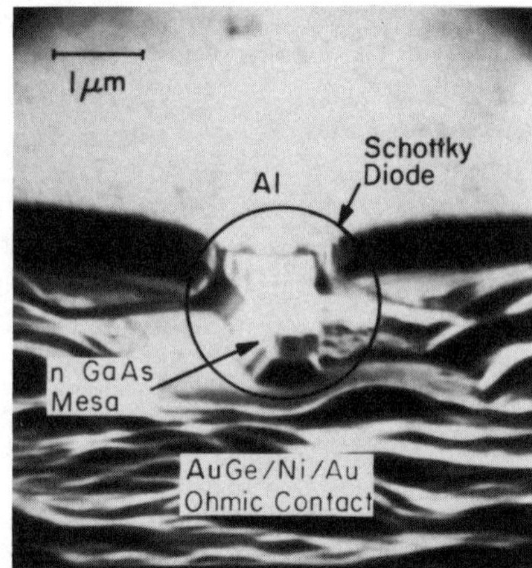

FIG. 33 (a) Imaging array with a planar Schottky diode at the apex of each bow-tie antenna. (b) SEM photographs of a planar Schottky diode (LIHS: side view; RHS: end view).

Mariner missions (P. D. Potter) where they were implemented to efficiently transmit *X*-band (8–12 GHz) and reflect *S*-band (2–4 GHz) radiation. Likewise, in a plasma experiment, a dichroic plate is utilized to pass the far-infrared radiation at the source frequency ω_o, but reflect radiation at the plasma and electron cyclotron frequencies which are $\simeq\frac{1}{3}\ \omega_o$. In

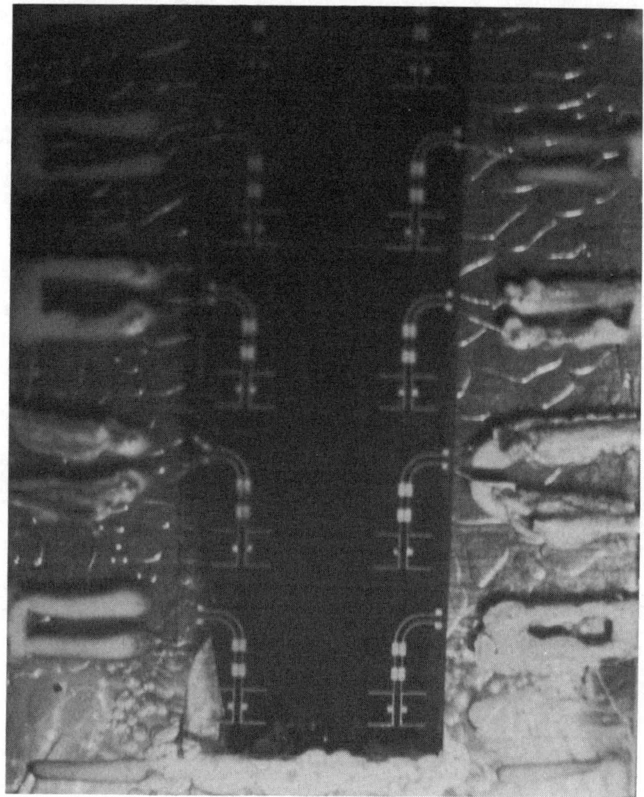

FIG. 34 Hybrid twin dipole imaging array with beam lead Schottky diodes.

effect, the dichroic plate acts as a high bandpass filter. The plate consists of a perforated aluminum sheet whose thickness is not negligible with respect to λ_o. This plate is mounted over the open aperture and designed to provide low losses plus minimal degradation to the beam pattern and polarization integrity. For a source at $\lambda_o = 1222$ μm, the plate thickness employed was 1200 μm, with holes of 850 μm diameter and spacing of $\simeq$900 μm. Typical source radiation losses measured were $\leq$10% with noise reduction of order $\simeq 10^2$.

For microturbulence investigations ($\omega/2\pi \leq 1$ MHz), the IF amplifier system consists of two low noise amplifiers with noise figure, N.F. $\simeq$ 1.8 dB and total gain, $G = 52$ dB. The upper 3 dB point of the gain bandwidth is $\simeq$10 MHz. The first section amplifier is high gain (40 dB), low noise to maximize the signal-to-noise ratio while the second stage (12 dB) serves to drive the cable line to the data room ($\simeq$40 m cable length) where it is terminated into 50 ohms. A capacitor on the amplifier output constitutes a high pass filter which is designed to provide a low frequency 3 dB cutoff point of ≃10 kHz. Such a system easily accomodates the frequency range of interest.

In the case of ICRF wave studies, the signal frequency is in the range of 10 − 100 MHz. Referring to Fig. 35, the amplified IF signal is split into two parts. One portion of the IF signal passes into a spectrum analyzer operated as a tuned narrowband amplifier (bandwidth ≃ 300 kHz). The output of the spectrum analyzer (which is proportional to the amplitude of the scattered electric field) is then fed into the tokamak data system. The second part of the IF signal is correlated with a reference from the ICRF antenna to generate a phase signal (ϕ_s) via a phase sensitive mixer. The envelope of this phase signal is equivalent to amplitude.

In the data room, the IF signal is digitized by a waveform digitizer (Le Croy 2264) capable of 4 MHz maximum sampling rate (2 MHz was generally used for these measurements). Available memory modules (Le Croy 8000/8) permit a maximum of 32 k words (8 bit) per channel for a six channel system. The digitized data is then handled by a DEC PDP-11/34 computer which provides the interface between the main computer (DEC VAX-11/780) and all other data collection devices. All data, for each shot, are permanently stored in the TEXT archive library.

To obtain the absolute density fluctuation amplitude of the plasma

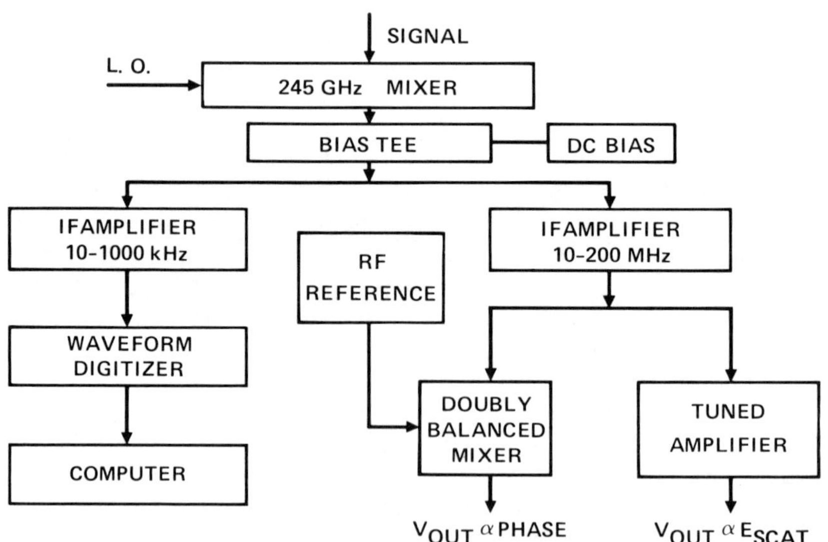

FIG. 35 Data processing system for ICRH measurements.

waves the receiver system, consisting of the Schottky diode mixer, bias tee, and IF amplifiers, must be calibrated. This is achieved via standard hot-cold load techniques and Y-factor analysis. The mixer noise temperature (T_M), the IF system noise temperature (T_{IF}) and the mixer conversion loss (L_{conv}) are measured. The conversion loss of the mixer results from the mismatch between the antenna impedance (50 Ω) and the RF impedance of the diode (≃200 Ω). Also, reflection losses from impedance mismatch between mixer and IF amplifier (50 Ω) must be considered.

In the hot-cold load technique, the noise power difference at the IF amplifier output is measured when the receiver (and separately, the IF amplifier) is alternately connected to a room temperature load and to a load cooled to liquid nitrogen temperature (77°K). For the measurement reported herein, Emmerson-Cumming Eccosorb was used as the absorber. Specifically, the receiver system noise temperature and mixer conversion loss are given by

$$T_s(DSB) = \frac{P_n(C)T_H - P_n(H)T_C}{P_n(H) - P_n(C)}, \tag{20}$$

and

$$L_s = \frac{T_H - T_C}{(Y_H - Y_C)(T_H + T_C)}\left(1 \pm \frac{L_s}{L_i}\right). \tag{21}$$

In the above T_s(DSB) is the double side band system noise temperature, $L_s(L_i)$ is the conversion loss at the signal(image) frequency, $T_H(T_C)$ is the temperature of the hot(cold) load, P_n is the detected noise power, and the Y-factors are defined as follows:

$$Y_H = \frac{P_n(H)}{P_n(IF)} \quad \text{and} \quad Y_C = \frac{P_n(C)}{P_n(IF)}, \tag{22}$$

where $P_n(\mathrm{IF}) = (T_H + T_C)k_B B G_{\mathrm{IF}}$. Here B is the IF bandwidth, G_{IF} is the gain, and k_B is Boltzmann's constant. The IF noise temperature is obtained by again using the hot-cold load technique. For these experiments, no attempt was made to enhance the mixer sensitivity at the signal frequency. Therefore, $L_s \simeq L_i$, and a double sideband measurement was employed.

Typical mixer noise temperature of 2500°K are measured with a total system noise temperature of 500°K. The IF amplifier noise temperature is ≃150°K and mixer conversion losses are ≃13 dB. For the raw data, a voltage $[V_s(t)]$ is measured corresponding to the scattered electric field $[\mathbf{E}_s(t)]$. The digitizer is terminated in R = 50 Ω. Therefore, if we know V_s, R and the conversion loss of the detection system, the total scattered power can be calculated ($P_s = V_s^2/R$).

With the above calibration procedure in mind, it is appropriate to perform a simple calculation to provide a feeling for the minimum detectable density fluctuation level. For this purpose, we assume $P_o \simeq 5$ mW, $\lambda_o \simeq 1$ mm, and $T_s = 5000°$K DSB. Then using Eq. (7) we find that

$$P_s = \frac{1}{4} P_o r_e^2 L_v^2 \lambda_o^2 \tilde{n}^2 \simeq 10^{-28}[\tilde{n}(cm^{-3})]^2 \text{ W}. \quad (23)$$

The major noise sources are the plasma synchrotron radiation background and the receiver noise level. To calculate their magnitude, $P = k_B T B$, we require the temperature (T) and the receiver bandwidth (B). Here let us assume we have launched a coherent wave and assume a bandwidth of $B \simeq 50$ kHz. Then, taking $T_e = 5$ keV and $B_o = 40$ kG and some chamber wall reflectivity (i.e., an effectively optically thick plasma), it is appropriate to assume a 5 keV blackbody in calculating the synchrotron radiation background. The resultant plasma and receiver noise background powers are 3.8×10^{-11} W and 3.4×10^{-15} W, respectively. Without recourse to sophisticated signal processing techniques, we still calculate a minimum detectable $\tilde{n}_{\min} \simeq 6 \times 10^8$ cm^{-3} (for a signal-to-noise ratio $S/N = 1$) or with $n_e = 10^{14}$ cm^{-3} we have $(\tilde{n}/n_e)_{\min} \simeq 10^{-15}$. In the case of a broad spectrum such as appropriate for drift wave type microturbulence (i.e., $B \simeq 10^6$ Hz), we still obtain $\tilde{n}_{\min} \simeq 3 \times 10^9$ cm^{-3}.

The absolute calibration of the density fluctuation amplitude has been directly verified experimentally by scattering from a launched plasma wave of known magnitude (as determined via Langmuir probes). The launched ion acoustic wave is unidirectional so k-matching conditions are met all across the scattering volume. The scattered signal was observed to have a signal-to-noise ratio of $\simeq 1.3$. Under k-matching conditions the predicted scattered signal can be determined from Eq. (7). The ratio of this result, P_s, to P_m (noise power of the mixer) produces a signal-to-noise ratio 1.6 ± 0.6, which is in good agreement with the measured value, thus validating the calibration procedure.

VI. Calibration of Optical System

For any scattering experiment, calibration of the optical system is essential. It is necessary to precisely know scattering parameters such as wavenumber resolution and scattering volume. Also, for a multichannel apparatus, system parameters including the relative collection efficiency, local oscillator power, and mixer sensitivities for each channel, are required in order to ascertain the absolute level of density fluctuations.

An acoustic cell (Park et al., 1982; Park et al., 1985; Gordon, 1966; Saito et al., 1981) is employed for optical system calibration. Dodel et al.

(1984) have also made use of a rotating wire instead of the acoustic cell following the method of Jones et al. (1984). As shown in Fig. 36, the acoustic cell consists of a main body possessing good transmission at the desired FIR operating frequency (e.g. TPX, polyethylene) and a piezoelectric transducer (PZT) to launch acoustic waves of the appropriate frequency. Unfortunately, even if the PZT driving voltage is kept constant the amplitude of the acoustic waves is a sensitive function of frequency. This variation arises since PZT transducers exhibit a series of resonances related to their thickness d by

$$d = \frac{\lambda}{2}, \frac{3\lambda}{2}, \frac{5\lambda}{2}, \ldots, \tag{24}$$

where λ is the acoustic wavelength in the PZT transducer. Because of this strong frequency dependence, the amplitude of the launched waves must be measured within the acoustic cell at each frequency. As shown in Fig. 36, this is accomplished by means of small-angle HeNe laser scattering orthogonal to the FIR scattering plane. To achieve this, TPX plastic ($5 \times 5 \times 15$ cm) is employed as the main body since it has good transmission in the FIR as well as adequate transmission in the visible. A 1.5 mW cw HeNe laser is employed as the probe source with detection accomplished through use of a photomultiplier tube. The resultant frequency responses are shown in Fig. 37 for two different PZT thicknesses ($d = 1.27$ cm, 2.2 cm) with the ac excitation voltage held constant at $\simeq 50$ V. The resonances are obvious. To provide flexibility in system calibration, the transducers are not permanently attached to the cell bodies but instead are simply clamped solidly to the body with a modicum of vacuum grease at the interface. This technique provides good acoustic coupling and facilitates the selection of varying PZT thicknesses for frequency selection.

The accuracy of the system calibration is (obviously) dependent on a precise knowledge of the acoustic wavelength within the cell. As shown in Fig. 38, this measurement is easily performed using a time-of-flight technique to determine the sound speed in the medium (and thus k since ω is known). Here, the PZT excitation voltage is no longer a steady sinusoid, but is instead gated using a tone burst generator. From the measurements shown in Fig. 38, the sound speed in TPX is observed to be $\simeq 2.2 \times 10^{15}$ cm/s. It should be noted that this time-of-flight technique was greatly facilitated by cutting a "sawtooth" pattern on the end of the cell body (see Fig. 36) to eliminate reflections and possible resonance problems. Basically, this simple pattern serves as a "beam dump". As a final point, it should be noted that the acoustic cell displays the same behaviour as a classical damped oscillator when driven on or off resonance. This is clearly illustrated in Fig. 39.

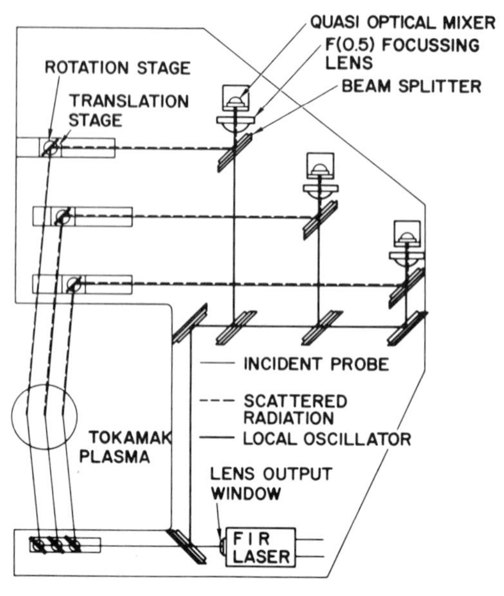

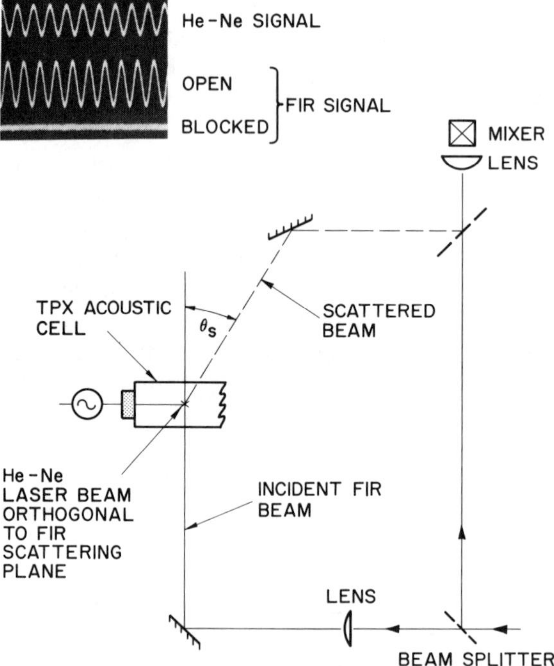

FIG. 36 Schematic of acoustic cell calibration scheme (Park et al., 1982).

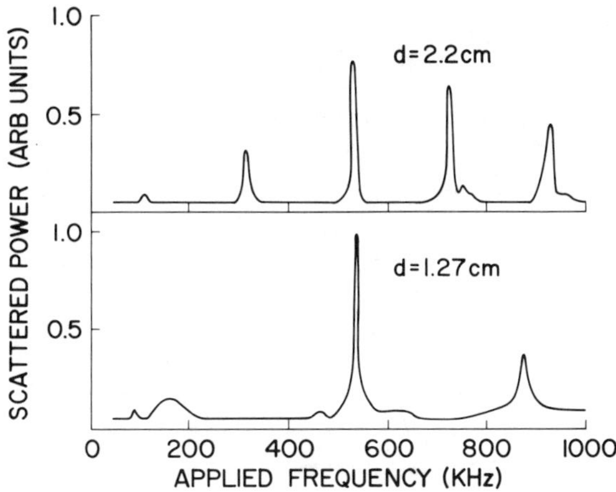

FIG. 37 Acoustic cell frequency response with two different thicknesses of piezoelectric transducers (Park et al., 1982).

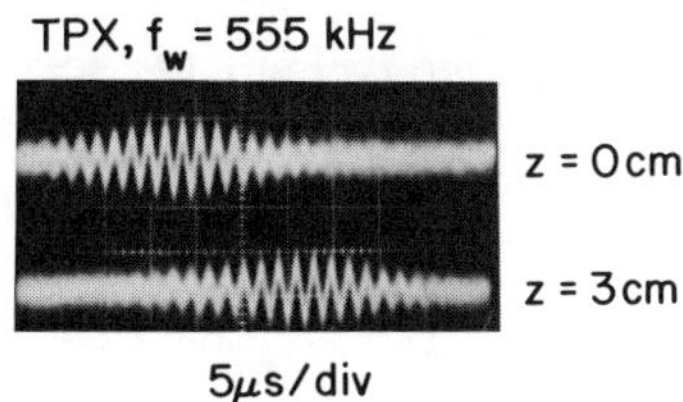

FIG. 38 Time-of-flight determination of acoustic speed (and hence wavenumber) in TPX acoustic cell (Park et al., 1982).

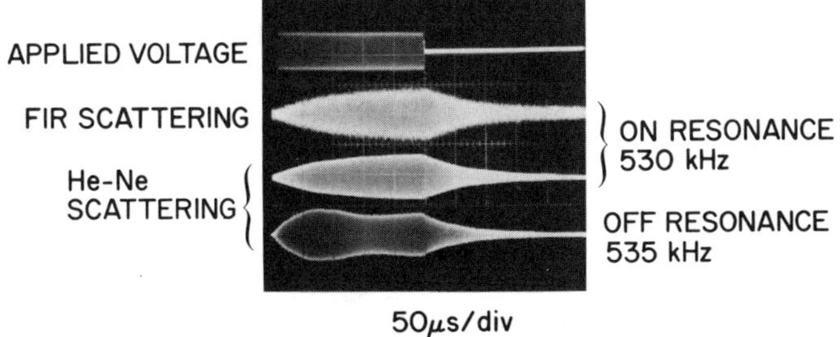

FIG. 39 FIR and HeNe scattering response from acoustic cell driven on and off resonance with a tone burst generator (Park et al., 1982).

Despite their convenience, acoustic cells do have some associated problems which must be noted. First, there is obviously a bending of the FIR radiation at the dielectric-air interface for all angles other than exact normal incidence. One must, therefore, employ Snell's law and correct the data for this angular shift. A second difficulty is associated with the PZT resonances described earlier. As will be seen in the following, the fourth scattering channel ($\theta_s = 13.5°$) on the Microtor apparatus has not yet been calibrated due to a lack of acoustic cell response at the appropriate frequencies (see Fig. 37).

Using the above described techniques the optical system may now be calibrated. First, the wavenumber resolution for each channel was determined. In this measurement, the acoustic frequency (and hence wavenumber) was varied while the acoustic wave amplitude was held constant (as monitored by HeNe laser scattering) and the scattered power monitored on the appropriate channel. The result is a plot for each channel as shown in Fig. 40. From these curves it is determined that the wavenumber resolution is $\Delta k \simeq 1\ \text{cm}^{-1}$, as measured at the electric field e^{-1} points. This result is in good agreement with the theoretically predicted value $\Delta k \simeq 2/a_o$ using the separately measured value $a_o \simeq 2$ cm for the beam waist.

The scattering length was determined by scanning the acoustic cell along the incident beam and monitoring the scattered power. The experimental results are shown in Fig. 41. The scattering length defined as the full width at half-maximum of these curves is in good agreement with the theoretical value of Eq. (11).

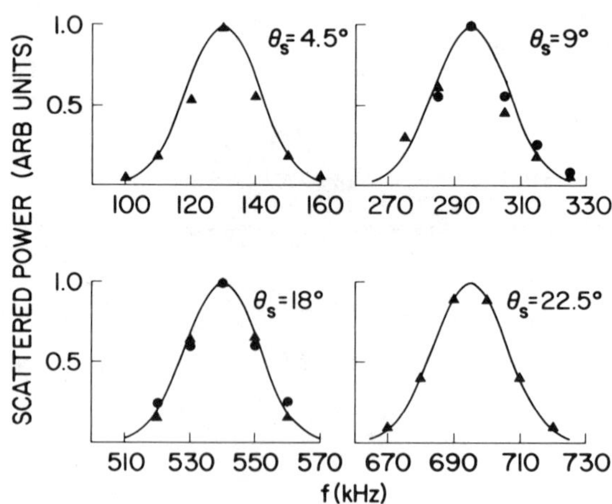

FIG. 40 Acoustic cell calibration of scattering wavenumber resolution (Park et al., 1982).

Similarly, the wavenumber resolution and scattering volume length were measured by Dodel et al. (1984) via use of a rotating wire scheme with a 199 μm scattering system. Here a thin wire (0.3 mm diameter, mounted on a rotating disc) passes across the laser beam and the Doppler shifted radiation diffracted by the wire is in turn detected. The transverse dimension of the scattering volume is shown in Figs. 42(a) and (b) for scattering angles of 8° and 2°, respectively. By using Eq. (9) and the measured scattering volume transverse dimension, the wavenumber resolution is determined. By translating the rotating wire along the laser beam, the length of the scattering volume may also be measured.

The final optical system parameter that must be obtained is the relative collection efficiency of the various channels. This may be determined by two methods. First, the signal level on each channel for a fixed acoustic wave amplitude is measured where the appropriate acoustic wavelength for k matching is selected for each channel. The results are shown in Fig. 43(a) for the multichannel apparatus on the Microtor tokamak. Another method, employed on the TEXT system, uses an arrangement of three mirrors to direct the FIR probe beam into a particular channel by simulating the scattering angle θ_s. A monitor of the beam is required to ensure that a constant power level is maintained. Then, using the exact optics (lens, beam splitter, etc.) to be employed in the actual scattering experiment, the signal level is measured for each channel. This measurement is made one channel at a time with the same detector and equivalent local oscillator power. The results are shown in Fig. 43(b). Within the error of the measurement, the relative collection efficiency of each channel is approximately unity.

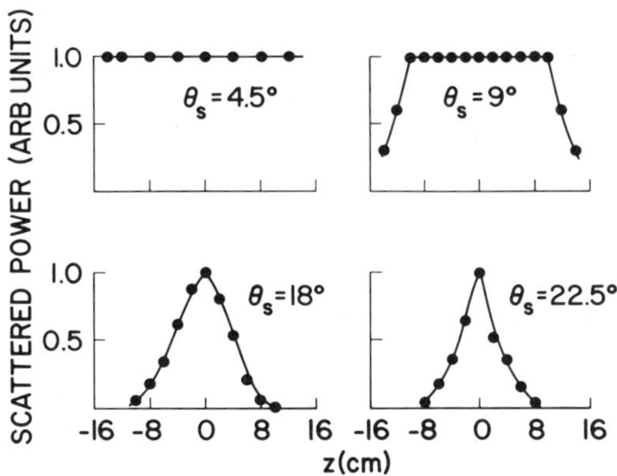

FIG. 41 Acoustic cell calibration of scattering volume length at four scattering angles (Park et al., 1982).

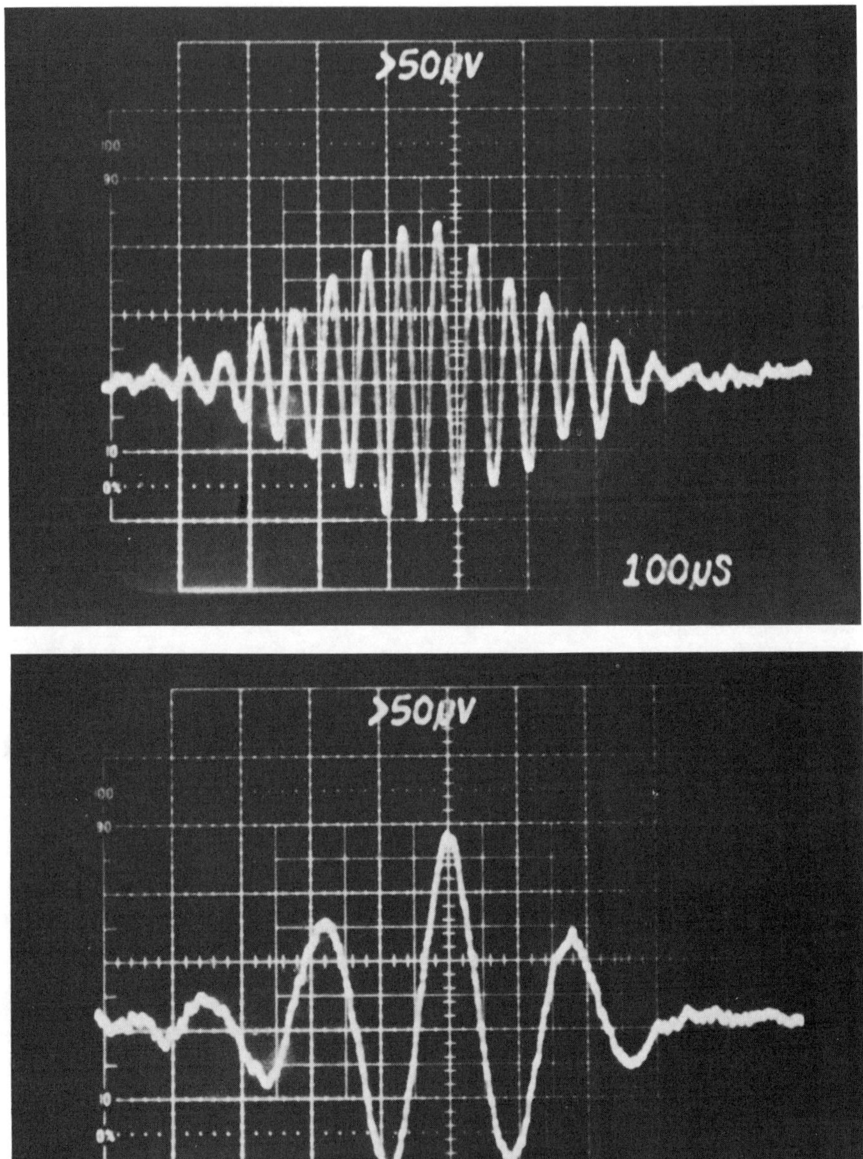

FIG. 42 Signals obtained by scanning the scattering volume with a rotating wire for (a) $\theta = 8°$, and (b) $\theta = 2°$. (One division on temporal axis $\simeq$ 1.6 mm) (Dodel et al., 1984)

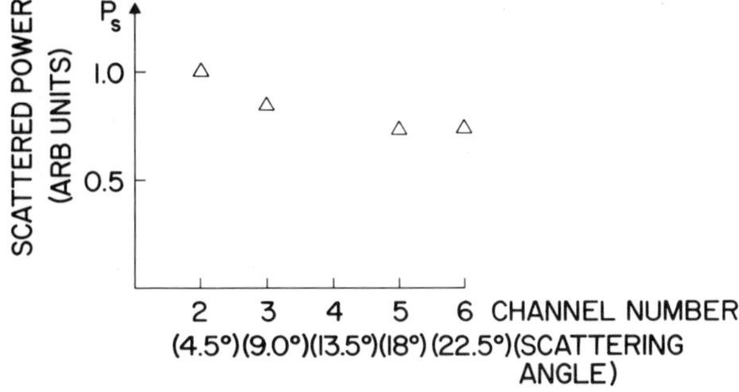

FIG. 43 Calibration of system collection efficiency; (a) using acoustic cell techniques with the system on Microtor (Park et al., 1982) and (b) using three configuration with the system on TEXT.

In an actual experiment, the system functions with all six channels simultaneously. Therefore, it is essential to know the local oscillator power provided to each channel along with the relative mixer sensitivities. As mentioned previously, the beam splitters chosen to divide the local oscillator beam give roughly the same RF power to each channel. To accurately determine this value a scheme described by Park et al. (1985) will be used. The same PZT transducer employed to drive an acoustic wave in the TPX cell is simply placed on top of the FIR laser cavity with the acoustic waves being launched perpendicular to the longitudinal axis. The resulting compression and rarefaction of the molecular gas medium (i.e. laser gain medium) results in an amplitude modulation of the FIR output. The oscillator used to drive the PZT is set to a frequency of interest, ω_m, for these measurements (say 100 kHz). Hence, a signal at ω_m is detected at the mixer.

The IF amplifier output signal (due to wave scattering) for each channel is:

$$P_{\text{out}_i} = G_i C_i (\alpha_i P_{LO} P_S)^{1/2} \qquad i = 1, 2, 3, \ldots 6. \tag{25}$$

where

G_i = IF amplifier gain

C_i = mixer sensitivity

α_i = assignment factor of local oscillator power.

For the relative calibration arrangement of the LO power there is no scattering signal. Therefore, the mixer output signal voltage is

$$P_{\text{out}_i} = G_i C_i \alpha_i P_{LO} \qquad i = 1, 2, 3, \ldots 6, \tag{26}$$

where the amplitude modulated P_{LO} is

$$\begin{aligned} P_{\text{LO}} &\propto [E_0 + E_m]^2 \\ &= [E_0 \cos(\omega_o t) + E_m \cos(\omega_m t) \cos(\omega_o t)]^2 \end{aligned} \tag{27}$$

In the above, E_0 and E_m are the unmodulated and modulated portions of the LO beam, respectively.

Since the IF amplifier is unable to respond to RF frequencies and $E_0 \gg E_m$, the amplifier output voltage is

$$P_{\text{out}_i} = G_i C_i \alpha_i E_m E_0 \cos \omega_m t, \tag{28}$$

Thus, from this measurement the relative value of $G_i C_i \alpha_i$ is found.

In order to determine the relative calibration factor, $G_i C_i (\alpha_i)^{1/2}$, it is necessary to measure the relative mixer sensitivities, C_i. To do this, the local oscillator power is kept constant for each channel (by using the same beam splitter) and then the amplifier output voltage is measured for each channel (one channel at a time). For such an arrangement the signal voltage is

$$P_{\text{out}_i} = G_i C_i \alpha_o E_m E_0 \cos \omega_m t \qquad i = 1, 2, 3, \ldots 6, \tag{29}$$

where α_o is the same for each channel since the LO is the same for each channel. From this measurement the relative value of $C_i G_i$ is determined.

Now by simply multiplying $G_i C_i \alpha_i$ from Eq. (28) and $G_i C_i$ from Eq. (27) and taking the square root we have the desired result $G_i C_i (\alpha_i)^{1/2}$.

With these calibration factors in hand, along with the absolute power measurement of the FIR laser and the detector conversion loss, it is possible to determine the absolute magnitude of the plasma density fluctuations.

VII. Typical Scattering Data

As mentioned previously, multichannel far-infrared scattering systems have been installed on the UCLA Microtor tokamak and the TEXT National User's Facility tokamak administered by the Fusion Research Center, University of Texas, Austin. The TEXT investigations have resulted in a detailed characterization of low-frequency tokamak microturbulence, while ICRF heating studies have dominated experimental efforts on Microtor. Typical scattering data for these two types of measurements will be shown in this section in order to demonstrate the performance potential for a multichannel FIR scattering system.

A. MICROTURBULENCE

A major obstacle impeding confinement in prototype thermonuclear fusion reactors results from instabilities which are manifested through bulk distortions of the plasma column (magnetohydrodynamic instabilities) and fine-scale microturbulence. In present day tokamaks, the experimentally determined electron energy confinement time is estimated to be two orders of magnitude worse than neoclassical predictions. Low-frequency microturbulence driven by plasma inhomogeneities (e.g. ∇n and ∇T, drift wave turbulence) is universally observed in tokamaks and generally considered to be related to the measured anomalous transport levels (Waltz et al., 1980; Horton, 1984; Rosenbluth and Rutherford, 1981). By employing collective scattering of far-infrared radiation, the spectrum and magnitude of spontaneously occurring tokamak microturbulence (i.e. density fluctuations, $\tilde{n}(\mathbf{k},\omega)$) has been thoroughly measured as a function of position in the TEXT (Brower, 1984; Brower et al., 1985a, 1985b, 1985c) and Microtor (Park et al., 1982; Park, 1984) devices. The turbulent nature of the phenomenon to be explored and the discharge irreproducibility dictate the need to determine the complete time-resolved frequency and wavenumber spectra during a single tokamak discharge. This type of measurement requires the use of a multichannel scattering apparatus.

Representative data from the TEXT tokamak are shown in Fig. 44, which gives the frequency spectra of the scattered power at six discrete wavenumbers measured simultaneously during a single discharge. The scattering volume is located along a vertical chord through plasma center and the spectra are thereby dominated by fluctuations with poloidal

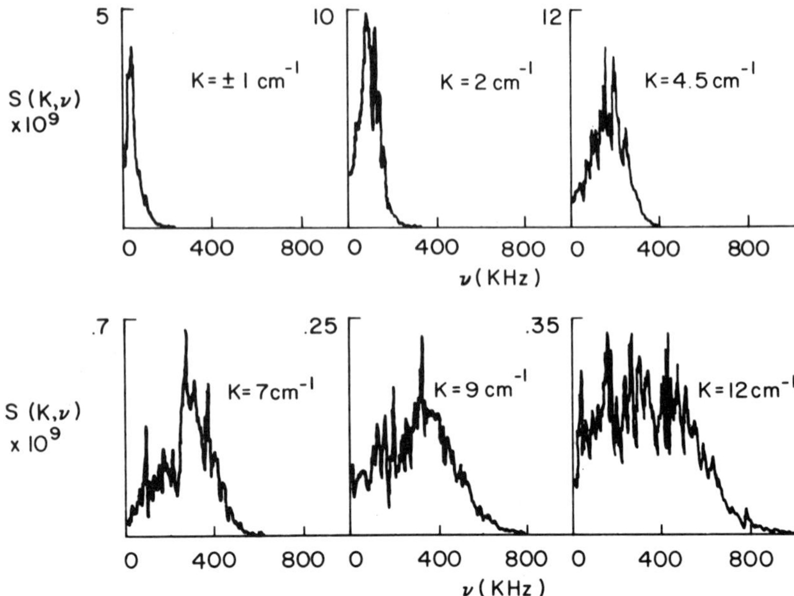

FIG. 44 Multichannel homodyne frequency spectra for poloidal fluctuations (Brower et al., 1985b).

wavenumber. The discharge conditions for TEXT ($R = 1$ m, limiter radius $a = 27$ cm) were $I_p = 300$ kA, $B_T = 2.8$ T, $\bar{n}_e = 3.5 \times 10^{13}$ cm^{-3}, $T_{eo} = 950$ eV, $T_{io} = 600$ eV, $Z_{\text{eff}} \simeq 2$, and $\tau_e \simeq 12$ ms. The frequency power spectra are produced by performing a Fourier transform analysis of the digital time series representation of the scattered electric field.

Four features are readily distinguished from the frequency spectra: (1) a distinct peak for a particular wavenumber, (2) a shift in the peak to higher frequencies as k_θ increases, (3) a broadening of the spectral linewidth with larger k_θ, and (4) a significant decrease in the magnitude of the fluctuation level for $k_\theta \geq 7$ cm^{-1}, as evidenced by the calibrated ordinate. For a particular k_θ, the average frequency peak was consistent from shot-to-shot although the fine structure was not reproducible. This demonstrates the need for a multichannel system.

Advanced analysis techniques used in the study of nonlinear three-wave coupling (Hasegawa and Kodoma, 1978; Mohamed-Benkadda et al., 1984; Ritz and Powers, 1985) also require a multichannel scattering scheme. In such a system, the scattered radiation collection optics are oriented such that $\mathbf{k}_1 + \mathbf{k}_2 = \mathbf{k}_3$ and the nonlinear coupling between the modes $(\mathbf{k}_1\ \omega_1)$ and $(\mathbf{k}_2, \omega_2)$ in the mode $(\mathbf{k}_3, \omega_3)$ where $\omega_1 + \omega_2 = \omega_3$ are

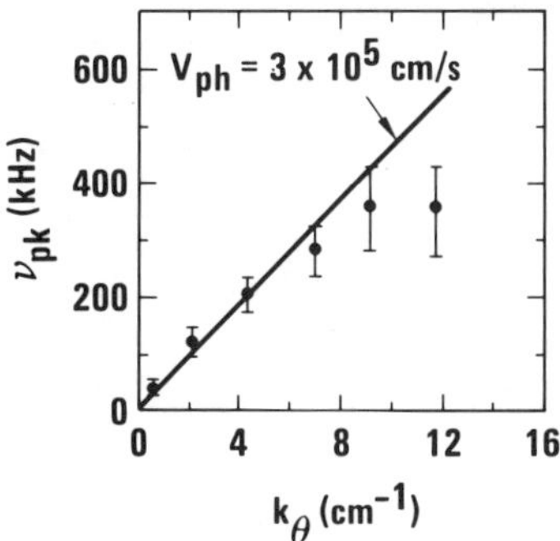

FIG. 45 Single shot statistical dispersion for k_θ density fluctuations. Note the distinct roll-over at large wavenumbers (Brower et al., 1985b).

examined. This technique has been employed to study three wave coupling in air nozzle turbulence by using a Rayleigh scattering technique (Mohamed-Benkadda et al., 1984). The nonlinear wave coupling coefficients are calculated by applying digital analysis methods such as the bispectra to examine the phase coherence (Ritz and Powers, 1985). The multichannel far-infrared scattering apparatus is easily reconfigured to comply with the wavenumber matching condition, making three-wave coupling measurements feasible. Preliminary measurements employing these techniques to identify nonlinear coupling in tokamak microturbulence are being carried out on TEXT (Ritz and Brower, private communication; 1985).

By plotting the peak frequency ($\omega_{pk}/2\pi = \nu_{pk}$) as a function of wavenumber (k_θ), a statistical dispersion relation may be produced as shown in Fig. 45. The measured dispersion demonstrates a linear frequency versus wavenumber dependence for $k_\theta \leq 9$ cm^{-1} with a phase velocity $v_{ph} \simeq 2$–$3 \times v_{De}$, the electron diamagnetic drift velocity. Heterodyne measurements, as shown in Fig. 46(a) where a homodyne spectroscopy technique is used (see Section IV.B), with the scattering volume situated at the plasma top, indicate that fluctuations are propagating largely in the electron diamagnetic drift direction, providing there is no Doppler shift due to plasma rotation. If the scattering volume is repositioned at the plasma bottom, again the fluctuations are seen to be propagating in the electron drift direction as shown in Fig. 46(c). The change in sign indicates a change in the wave vector direction of the fluctuation with respect to the collection optics, which is expected for waves traveling poloidally around the plasma cross section. When the scattering volume is positioned at the plasma center (see Fig. 46(b)), one scatters from fluctuations both above and below the midplane, thereby producing peaks in the frequency at $\pm\omega$. The shifts in sign from $+\omega$ to $-\omega$ are due to the momentum and energy conservation laws discussed in Section II. From Fig. 46 it is seen that the homodyne spectroscopy technique described in Section IV.B is indeed a sensitive one.

Similarly, by scattering along a vertical chord through the outside plasma edge, the measured fluctuations consist primarily of a radial wavenumber component, k_r. The single-shot frequency spectra for radial wavenumber fluctuations are shown in Fig. 47. Here no dispersive characteristics are observed, although the spectral linewidth continues to broaden with increasing k_r.

The wavenumber spectrum, obtained by integrating the frequency spectrum over all ω at each $k_\perp$, is shown in Fig. 48 ($t = -1$ ms), for poloidal fluctuations. It is characterized by a broad peak at $k_\theta \rho_s \simeq 0.25$ (ρ_s is the ion Larmor radius at T_e). The plasma parameters and profiles may be perturbed by injecting impurities. These perturbations are of interest in order to ascertain the scaling of the microturbulence with density and temperature profile changes, since the electron diamagnetic drift velocity, $v_{De} = T_e/eB(1/L_n)$ where $1/L_n = n^{-1}(dn/dr)$ is the density scale length. In addition, from theoretical analysis it is predicted that the magnitude of the density fluctuations is driven by the gradients in plasma temperature and density. Therefore, any correlation between changes in the density fluctuation level and the plasma profiles would serve as evidence that the source of the density fluctuations is drift wave type turbulence. The profile modifications are conveniently achieved via laser ablation of scandium into the plasma. The quantity of scandium injected is insufficient to significantly affect the plasma Z_{eff}; however, the neutral scandium atoms serve as a source of electrons which in turn produce substantial, time dependent changes in the plasma electron density. The injection can cause an increase in the centerline density of $\simeq$15%. The density profile typically increases at the edges first, where the scandium is initially ionized, and then the perturbation propagates towards the plasma center. This in turn results in considerable alterations to the density scale length. The time evolution of the wavenumber spectra during the impurity injection is shown in Fig. 48. Promptly after the injection ($t = +1$ ms), the wavenumber spectra shifts to lower k_θ, then the total fluctuation level decreases to a minimum approximately 4 ms later. Full recovery of the wavenumber spectra is observed in $\simeq$12 ms. The time evolution of $\tilde{n}(k_\perp, \omega)$ generally corresponds to changes in the density scale length. Initially, L_n increases with a similar reduction observed in the measured phase velocity ($v \propto 1/L_n$). The ensuing reduction in the total density fluctuation level may be attributed to a broadening of the density profile. Eventually, both the density profile and microturbulence spectra fully recover to their orig-

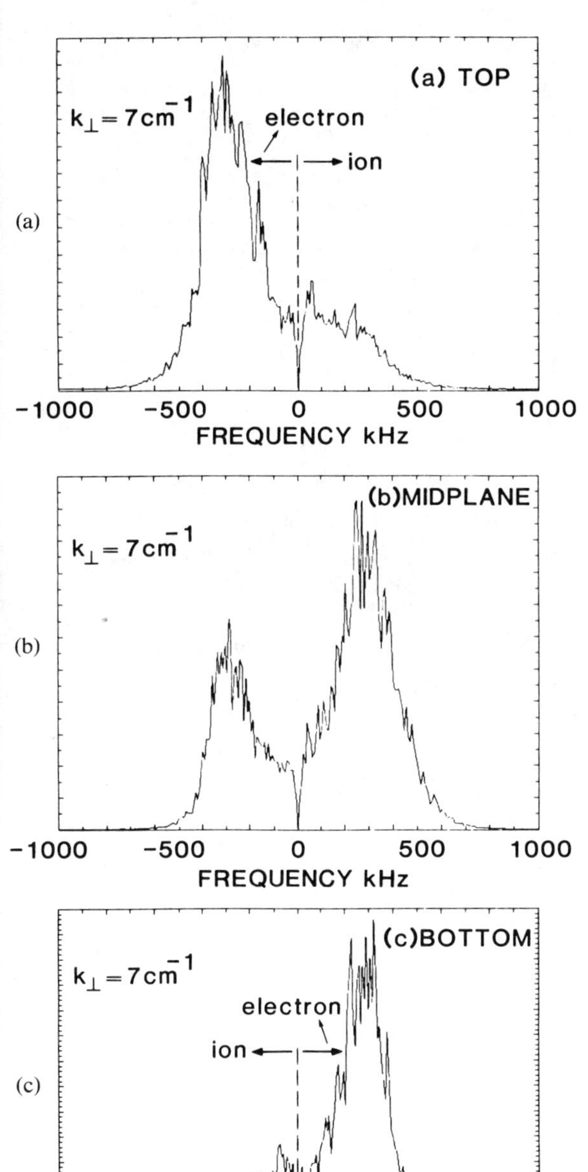

FIG. 46 Frequency spectra obtained using the homodyne spectroscopy technique; (a) plasma top, (b) plasma midplane, and (c) plasma bottom.

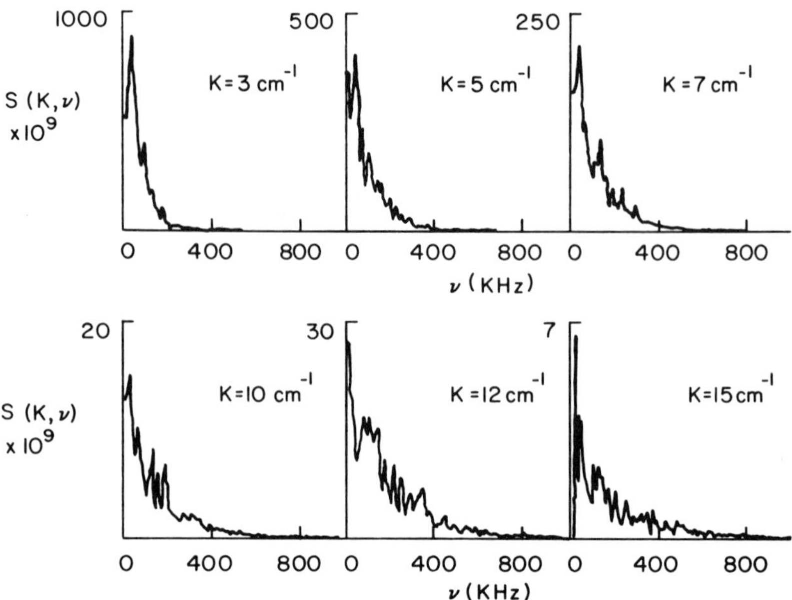

FIG. 47 Multichannel homodyne frequency spectra for radial fluctuations.

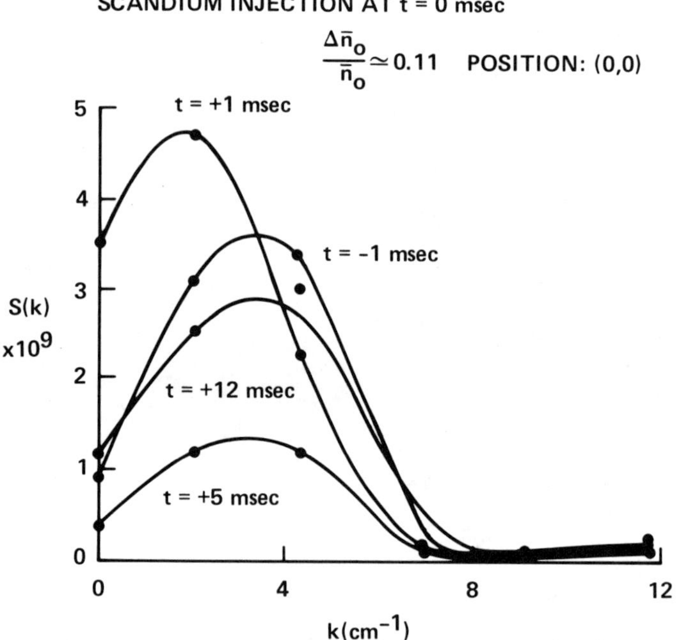

FIG. 48 Time evolution of the wavenumber spectrum during a perturbation resulting from scandium injection.

inal form. This correlation between $\tilde{n}$ and $n_e(r)$ establishes a direct link between plasma density fluctuations and drift wave turbulence. The ability to monitor the time evolution of the complete $k_\perp$ spectra is only possible with a multichannel system.

By scanning the scattering volume along a vertical chord through plasma center the spatial distribution of poloidal fluctuations may be determined. The center of the scattering volume is translated beyond the plasma edge so that the spatial resolution (L_v) is confirmed in situ by observing the decay of the scattered signal. Figures 49 and 50 illustrate the spatial distribution of the integrated scattered power for $k_\theta = 9$ and 12 cm^{-1}. From the plasma edge, $r/a = 1$, the scattered power drops to its e^{-1} point over 10(12) cm for $k_\theta = 12(9)\ \text{cm}^{-1}$. This implies a scattering volume length $L_v \simeq 20(24)$ cm; in agreement with Eq. (11) and the acoustic cell calibration results of Section VI. The asymmetric up-down spatial distribution of $P_s(k_\theta)$ is a new observation which is under continuing investigation (Brower et al., 1985a, 1985c). At present, it is unclear whether the observed asymmetry uncovers a fundamental physics question for tokamaks or whether it is the result of a machine-induced effect peculiar to the TEXT device.

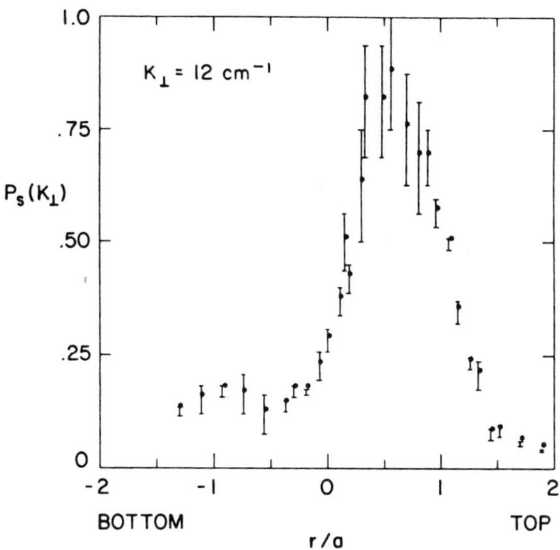

FIG. 49 Scattered power versus up-down position along a vertical chord at R for $k_\theta = 12\ \text{cm}^{-1}$.

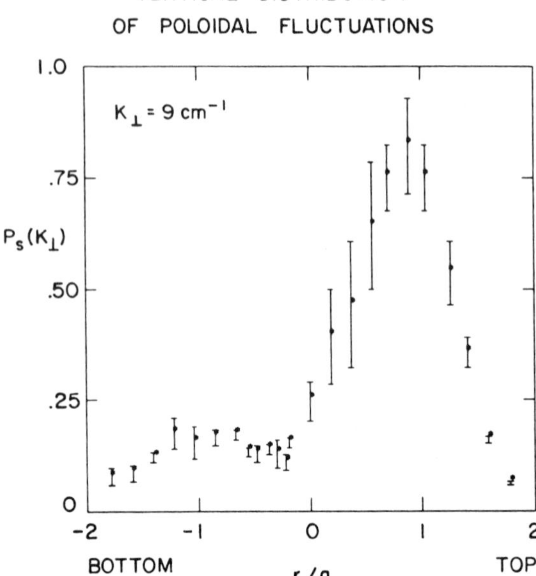

FIG. 50 Scattered power versus up-down position along a vertical chord at R for k_θ = 9 cm^{-1}.

In addition to the up–down asymmetry discussed above, other collective Thomson scattering measurements (Lipschultz et al., 1983; Slusher et al., 1984; Brower et al., 1985a) have noted substantial deviations from the standard observations of broadband ($\Delta\omega/\omega \simeq 1$) isotropic turbulence (Surko et al., 1983) in the plane perpendicular to the toroidal magnetic field. For example, although density fluctuations are commonly found to peak at the plasma periphery, a new tokamak edge phenomenon known as the 'MARFE' has been observed in the Alcator C tokamak (Lipschultz et al., 1983). This condition, which exhibits a density threshold for onset, is distinguished from ordinary turbulence by greatly increased radiation, density and density fluctuations, and decreased temperature in a small volume at the inner major radius edge of the plasma. In addition, a quasi-coherent fluctuation ($\Delta\omega/\omega \simeq 0.1$; Slusher et al., 1984) was recently observed during the H-mode regime of neutral beam heated PDX tokamak discharges. The narrowband fluctuation was noted as a temporal burst over a narrow region of minor radius near the separatrix. It should also be noted that all the above mentioned phenomena are associated with the poor confinement edge region of tokamaks.

The horizontal translation capability of the scattering volume on TEXT

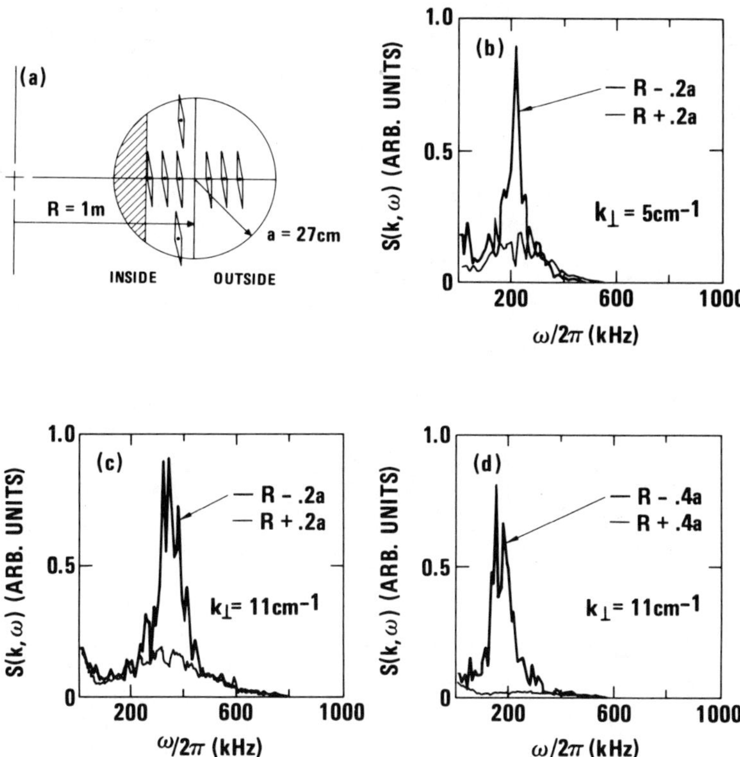

FIG. 51 Comparison of the fluctuation spectra between the torus inside and outside. (a) Scattering volume geometry for midplane positions $R \pm 0.2$a, $R \pm 0.4$a, $R \pm 0.6$a and vertical positions ± 20 cm above and below the midplane at $R - 0.2$a for $k_{\perp} = 11\ \text{cm}^{-1}$; the crosshatched area indicates plasma region not probed due to limited access ports. (b) $k_{\perp} = 5\ \text{cm}^{-1}$, $R \pm 0.2$a, (c) $k_{\perp} = 11\ \text{cm}^{-1}$, $R \pm 0.2$a, and (d) $k_{\perp} = 11\ \text{cm}^{-1}$, $R \pm 0.4$a (Brower et al., 1985c).

permits comparison of the density fluctuations between the high field torus inside and the low field torus outside. It was found that both the spectra and the magnitude of the fluctuations varied significantly at in–out spatial positions centered on the equatorial midplane of the torus as shown in Fig. 51. The fluctuations on the torus inside are characterized by their large amplitude, quasicoherent (as low as $\Delta\omega/\omega \simeq 0.1$) nature when compared to the ubiquitous broadband microturbulence. Similarly, the wavenumber spectra of quasicoherent fluctuations were also found to be considerably narrower in $k_{\perp}$ as shown in Fig. 52. In addition, spatial scans along a vertical chord on the torus inside (as an example, see Fig. 51(a) at $R - 0.2$a) indicate that the narrowband fluctuations are only seen at the

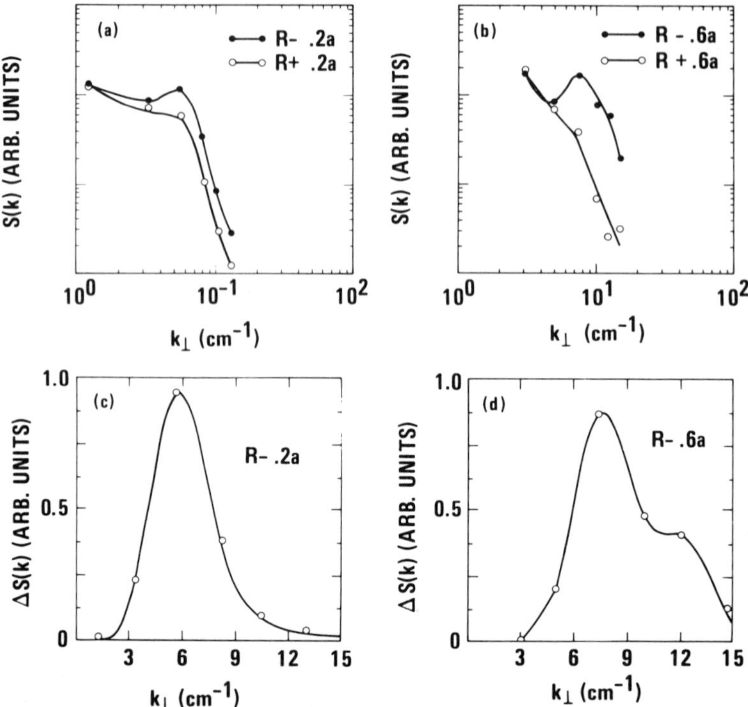

FIG. 52 Wavenumber spectra comparison at spatial positions (a) $R \pm 0.2$a and (b) $R \pm 0.6$a, with symbols ●, ○ denoting inside, outside, respectively. Wavenumber spectra for the quasicoherent fluctuation, $\Delta S(k_\perp) = S_{\text{in}}(k_\perp) - S_{\text{out}}(k_\perp)$, at positions (c) $R - 0.2$a and (d) $R - 0.6$a (Brower et al., 1985c).

plasma interior and are not observed at the edge. The spatially varying toroidal field ($B_T \propto 1/R$) is a likely source for this type of asymmetry, possibly resulting in a form of the toroidicity-induced drift wave (Chen and Cheng, 1980; Similon and Diamond, 1984). The internal localization of the quasicoherent fluctuation provides information on turbulence and transport in the central region of the tokamak plasma where global confinement is thought to be determined. The vertical and horizontal scanning capability of the apparatus are instrumental in providing several interesting experimental results.

B. ICRH Measurements

In order to achieve ignition temperatures in a fusion plasma, a variety of supplementary heating schemes have been under active study over the last few years (Hwang et al., 1983; TFR Group et al., 1982). Radio frequency heating in the ion cyclotron range of frequencies (ICRF) has proven one of the most effective methods, and future experiments are planned for TFCX, JET, TFTR, 'Big Dee', etc., although there remains a number of ambiguities. For example, the detailed mechanisms responsible for wave damping in different heating regimes are not fully understood. In addition, the associated problem of wave accessibility and propagation are also not generally monitored experimentally. Collective Thomson scattering has recently proved a useful and elegant technique in observing the short wavelength electrostatic waves resulting from, for example, mode conversion processes (A. Y. Lee et al., 1982; Park et al., 1984; TFR Group et al., 1982). This has served to improve the understanding of ICRF wave physics while also offering diagnostic applications (A. Y. Lee et al., 1982). The relevant Microtor parameters for data to be shown are B_T = 15–24 kG, I_p = 100kA, T_e = 500 eV, $T_i \simeq 200$ eV, $a = 10$ cm, $R = 30$ cm, and $\bar{n}_e \leq 5 \times 10^{13}$ cm^{-3} with a pulse duration of 15–25 ms. In addition to the multichannel scattering apparatus, there is a single channel system located 180° toroidally around the tokamak so that toroidal propagation effects can be investigated.

The experiments on Microtor have been quite successful and represent the first direct observation in a tokamak of externally excited electrostatic modes in ICRF (A. Y. Lee et al., 1982, Park et al., 1984). In the Microtor tokamak, the toroidal field varies both spatially and temporally. Figure 53 illustrates the temporal variation together with typical FIR scattering signals. These data were obtained in a deuterium plasma possessing a small hydrogen contraction ($\leq$10%). The applied rf was $\simeq$20 MHz, which is approximately the second harmonic frequency for deuterium. The observed signal changes are primarily caused by the time varying toroidal field. This sweeps ω/ω_{ci} and, therefore, modifies k_ω in a manner determined by the wave dispersion. The spatial variation in the FIR scattering signal is seen in Fig. 54, where signals are shown for the case of deuterium resonance (R = 25 cm) and hydrogen resonance (R = 32 cm) for the kinetic Alfven wave. From the time history of the spatial dependence of the scattering signals the wave dispersion relations can be obtained. Typical dispersion data are shown in Fig. 55, and compared with theoretical calculations.

One of the most important measurement needs in a magnetically confined fusion plasma is an accurate, time-resolved determination of the ion temperature distribution. At present, one of the most commonly utilized techniques for ion temperature determination is the energy analysis of charge exchange neutrals which escape from the interior of the plasma volume (Nexsen et al., 1979; Davis et al., 1979; Medley, 1981; Kaita and

SCATTERING SIGNALS

H^+, D^+ PLASMA
B_T = 22 kG PEAK(CENTER)
RF = 20 MH_z

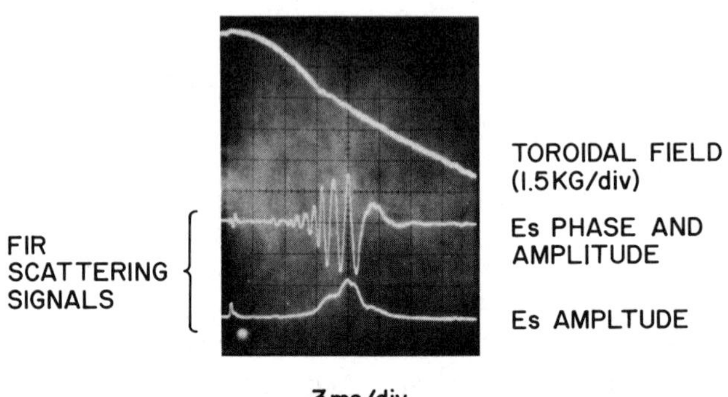

FIG. 53 Typical cw FIR scattering data from ICRF waves in the UCLA Microtor tokamak. As shown in the figure, the toroidal magnetic field varies temporally as well as spatially. At a fixed spatial position this sweeps ω/ω_{ci} in time and, therefore, modifies k_ω in a manner determined by the wave dispersion. This gives rise to time-varying phase and electric field amplitude signals determined by wavenumber matching and the wave properties (Luhmann and Peebles, 1984).

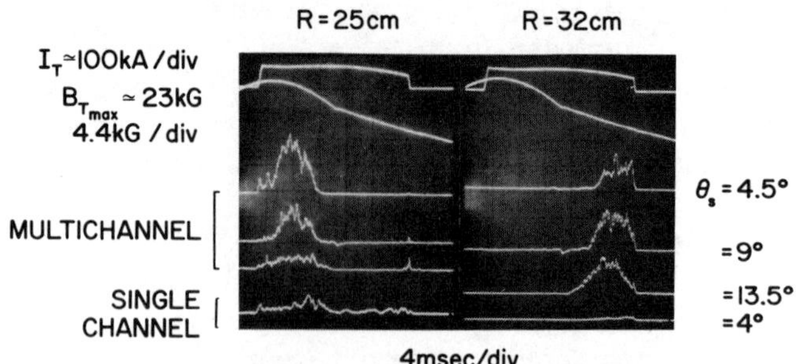

FIG. 54 Scattering signals from kinetic Alfven waves produced during ICRF heating in the Microtor tokamak (Park et al., 1982).

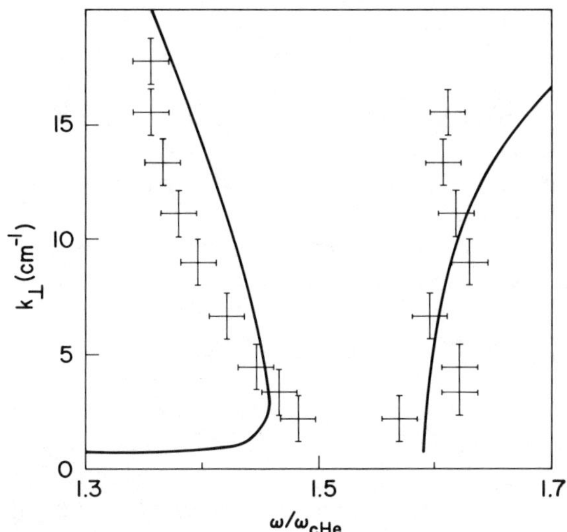

FIG. 55 Experimental points obtained via cw FIR laser scattering dispersion for both the kinetic Alfven and ion Bernstein waves in a H/He plasma in the UCLA Microtor tokamak (Luhmann and Peebles, 1984).

Medley, 1979; Medley and Persing, 1981; Sindoni and Wharton, 1978; TFR Group, 1978). However, there are several difficulties with this technique. First, the flux of escaping neutral atoms is considerably attenuated in passing through the plasma. Therefore, the measurement tends to be weighed by the cooler, outer regions of the plasma. An additional problem with this diagnostic is that the neutral atomic density peaks toward the outside of the plasma, further weighting the low-energy portion of the spectrum. Although these difficulties can be ameliorated by the use of active neutral beam probing (TFR Group, 1978) there remains interest in the use of alternative (or complementary) techniques. For example, the measurement of the Doppler broadening of spectral lines from high-Z impurity ions (Suchewer and Hinnov, 1978; Bitter et al., 1982) yields the ion temperature.

As mirror and tokamak plasmas become hotter and denser, the thermonuclear neutron yield also increases. Therefore, there is currently considerable effort aimed at the determination of ion temperature through a measurement of the neutron energy spectrum (Strachen et al., 1979; Hunt et al., 1980; Slaughter and Pickles, 1979; Slaugher, 1982). However, there are several processes which may confuse these measurements. Specifically, neutrons can be produced by reactions other than D-T and D-D. For example, the presence of high-energy electrons can result in neutron production both through direct dissociation of deuterium and through photodissociation of linear materials by the accompanying bremsstrahlung emission.

There have also been efforts aimed at the determination of fusion plasma ion temperature by means of collective Thomson scattering from *thermal* fluctuations (Bretz and Taylor, 1982; Kasparek et al., 1980; Kasparek and Holzhauer, 1981; Peebles and Herbst, 1978; Woskoboinikow et al., 1983). In this case, the shape of the scattering form factor $S(k)$ depends upon the ion temperature (albeit also on T_e and the effective impurity concentration). A major problem with this technique is that it requires a high-power pulsed laser source (i.e. megawatt power levels), which makes it difficult to provide time-resolved ion temperature profiles. In addition, since the post-detection signal-to-noise ratio $S = [s/(1 + s)](1 + B\tau)^{1/2}$, long laser pulses are required. Here S is the predetection signal-to-noise ratio, B is the bandwidth (or resolution) of each IF filter channel and τ is the detector integration time (limited obviously by the laser pulse duration). Therefore, given current CO_2 and FIR laser technology, reported signal-to-noise ratios have remained ≤ 3. In addition, the existence of nonthermal fluctuations and small impurity concentrations can mask or distort the ion feature. It is generally concluded that under optimum conditions the ion temperature can be determined to $\pm 15\%$ using this technique. However, to date there has not yet been a definitive thermal scattering determination of the ion temperature in a tokamak plasma.

An alternative measurement technique is to externally launch a plasma wave whose dispersion relation depends primarily on T_i and to measure its dispersion in a nonperturbing fashion via collective Thomson scattering. The unneutralized ion Bernstein wave (P. Lee et al., 1982a, 1982b; Wurden et al., 1982; Wurden, 1982; Schmitt, 1973; Schmitt and Krumm, 1976) represents a particularly good choice of target wave. The dispersion of this wave is relatively insensitive to electron density and temperature as well as parallel wavenumber. In addition, this mode possesses a strong electrostatic nature so that there are large associated density fluctuations which maximize the scattered power. Furthermore, for such a coherent mode the scattered power is greatly enhanced (over thermal) permitting the use of low-power cw lasers and multichannel systems. The major problem is in devising an excitation scheme which is compatible with fusion reactor plasmas. This ion temperature measurement technique is not being suggested as a replacement for the other techniques discussed above but rather as a complementary method for measuring such an important parameter as T_i.

The excitation technique is mode conversion of an externally launched fast Alfven wave. A fit of the experimentally determined ion Bernstein wave dispersion to the ion temperature dependent theoretical dispersion yields the local ion temperature as shown in Fig. 56. The partial ion temperature profiles (chord averaged) shown in Fig. 57, were obtained with temperature values consistent with charge exchange measurements. A multichannel system is required in order to produce the complete dispersion relationship for a particular spatial position during a single tokamak discharge.

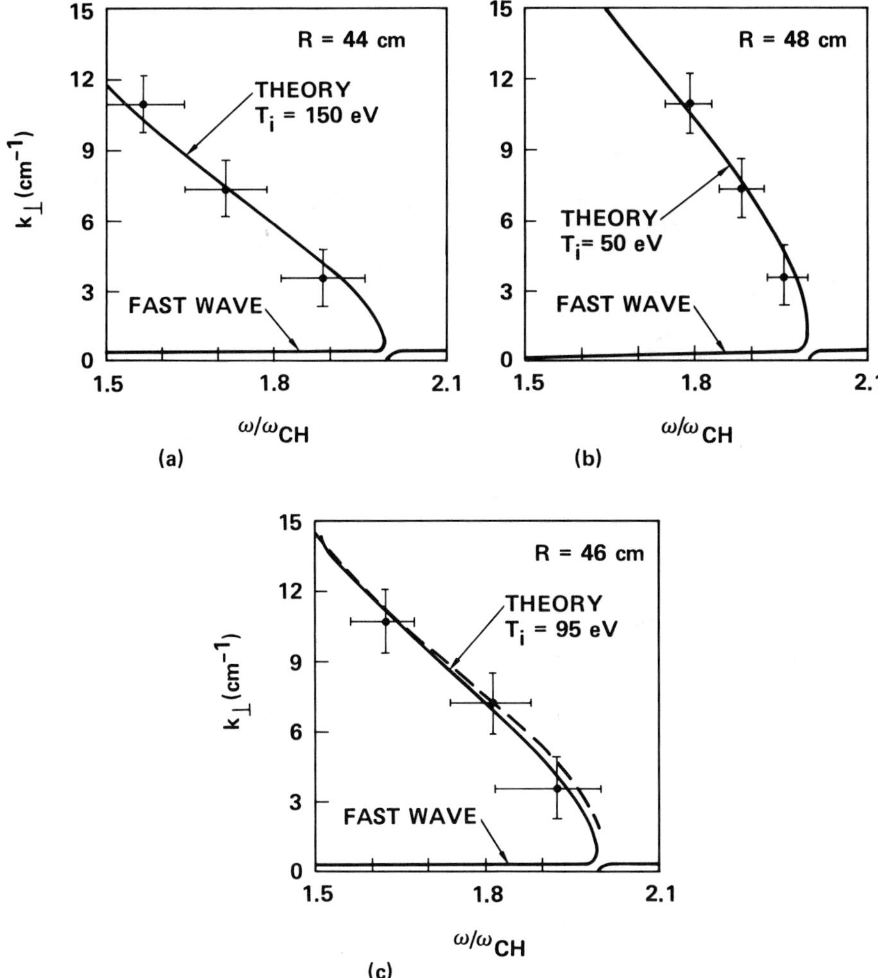

FIG. 56 Ion Bernstein wave dispersion relations at (a) R = 44 cm, (b) R = 48 cm, and (c) R = 46 cm.

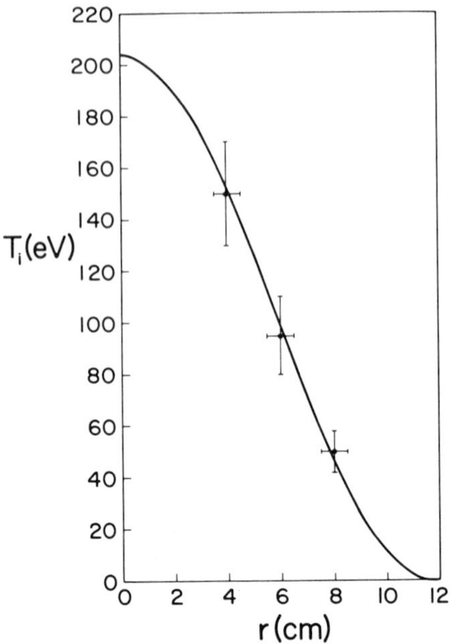

FIG. 57 Ion temperature profile determined from FIR scattering from ion Bernstein waves in the UCLA Microtor tokamak.

In the near future, the multichannel scattering apparatus on TEXT will be employed to scatter waves resulting from externally launched heating waves in the electron cyclotron range of frequencies (ECRF). ECRH systems have several applications in fusion plasmas including electron heating, profile control (use in α-particle heated plasmas), current drive and preionization (reduction of loop voltage) in tokamaks. Collective Thomson scattering will be of importance in order to fully understand the physics of ECRH prior to installation on a major fusion device and also to assess any deleterious effects on confinement as it pertains to microturbulence.

VIII. Other Applications

The application of the multichannel far-infrared scattering system to the study of basic tokamak plasma physics was briefly touched upon in Section VII. Through relatively minor reconfiguration of the detection optics, the diagnostic applications of such a system may be further expanded to

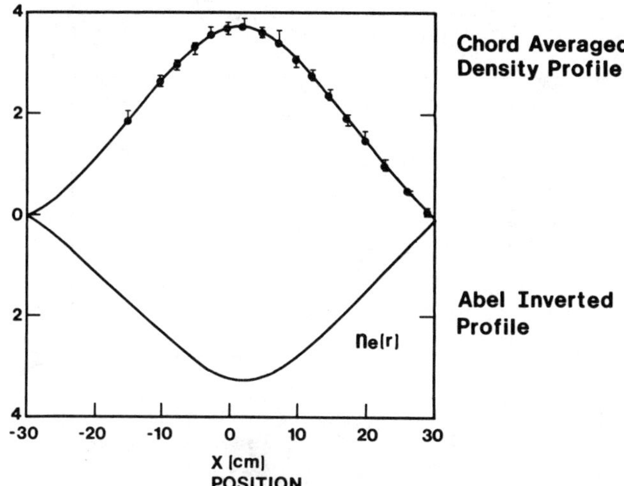

FIG. 58 Chord-averaged and Abel inverted density profiles for the TEXT tokamak obtained by using the FIR scattering system in the single-channel interferometer configuration.

provide electron density profiles via interferometry and poloidal field (current density) profiles via Faraday rotation measurements.

As mentioned in Section III, the 0° scattering channel may be employed as an interferometer to measure the plasma electron density. For a single chord measurement, the TEXT density profile is shown in Fig. 58, where 17 chords were measured on successive shots. Abel inversion of the chord-averaged results produces the local electron density profile shown in Fig. 58. The TEXT system can be expanded to a multichord measurement as discussed in Section IV.C, or imaging arrays may be employed, as on the Microtor tokamak (Young, 1984), with beam expansion techniques to produce a 20 channel measurement as shown schematically in Fig. 59.

Faraday rotation of the linearly polarized laser beam as it propagates through a tokamak plasma may be used to provide a chord-averaged measurement of the poloidal magnetic field profile, and thus infer the toroidal current density profile. Methods of Dodel and Kuntz (1978) and Soltwisch (1983) are easily adaptable to the FIR scattering system. The same imaging arrays used to measure the density profile may also be employed to determine the Faraday rotation angle, thereby providing $n_e(r)$ and $I_p(r)$ profiles for a particular tokamak discharge. These measurements are at present being actively pursued on the UCLA Microtor tokamak.

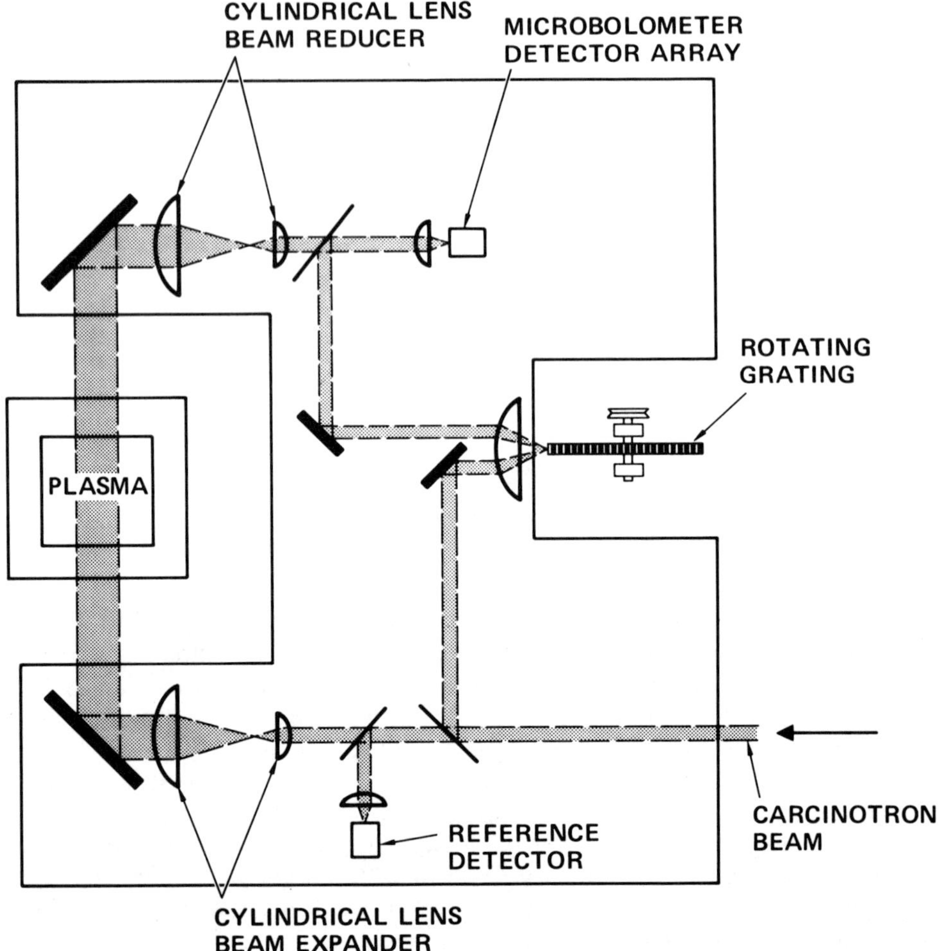

FIG. 59 Schematic of the simultaneous, 20-channel interferometer system on the Microtor tokamak.

ACKNOWLEDGMENTS

Design, construction and calibration of the multichannel scattering systems were greatly benefited by the assistance of R. L. Savage, Jr., C. X. Yu and D. Jungwirth. Through the course of the preparation of this chapter, the authors received important contributions from a number of individuals who provided both information and figures. They include D. Baker, G. Dodel, H. R. Fetterman, D. T. Hodges, D. B. Rutledge, A. Semet, T. Tsukishima, D. Vernon, M. Yamanaka, P. Young, and E. Holzhauer. Any errors or misinterpretations are the responsibility of the present authors and not the contributors. Tokamak machine time and operation were generously made available by the Fusion Research Center, University of Texas at Austin (TEXT) and the University of California at Los Angeles (Microtor). This work was supported by the U.S. Department of Energy, Office of Fusion Energy, under Contract Nos. DE-AM03-76SF00010 and DE-AC05-78ET53043 through subcontract No. UT-1-24120-52019.

REFERENCES

Apollo Lasers Inc, Dr. A. Semet, 9201 Independence Avenue, Chatsworth, CA 91311, private communication.

Archer, J. W., B. B. Greggor, R. J. Mattauch, J. D. Oliver (1982). *Microwaves,* **21,** 84.

Archer, J. W. (1984a). *IEEE Trans. Microwave Theory Tech.,* **MTT-32,** no. 4, pp. 416–420.

Archer, J. W. (1984b). *IEEE Trans. Microwave Theory Tech.,* **MTT-32,** no. 4, pp. 421–427.

Archer, J. W., M. T. Faber (1985). *IEEE Trans. Microwave Theory Tech.,* **MTT-33,** 533.

Asada, O., A. Inoue, and T. Tsukishima (1980). *Rev. Sci. Instrum.,* **51,** 1308.

Asada, O., K. Yoshioka, A. Inoue, and T. Tsukishima (1981). *Japan J. Appl. Phys.,* **20,** 173.

Baker, D. R., and S. T. Lee (1978). *Rev. Sci. Instrum.* **49,** 919.

Baker, D. R. (1980). *Rev. Sci. Instrum.,* **51,** 1304.

Belland, P., D. Veron and L. B. Whitbourn (1976). *Appl. Optics,* **15,** 3047.

Belland, P., and J. P. Crenn (1979). *Appl. Optics,* **18,** 1513.

Belland, P., and D. Veron (1980). *IEEE Jour. of Quantum Elect.,* **QE-16,** 885.

Belland, P. (1982). *Appl. Phys.,* **B27,** 123.

Bezzerides, B., D. F. Dubois, D. W. Forslund (1974). *Phys. Rev. Lett.,* **33,** 886.

Biegel, L., and T. L. White (1980). *Bull. Am. Phys. Soc.,* **25,** 891.

Bitter, M., S. Von Goeler, M. Goldman, K. W. Hill, R. Horton, W. Roney, N. Sauthoff, W. Stodiek (1982). Measurements of the central ion and electron temperature of tokamak plasmas from the X-ray line radiation of high Z impurity ions, *Sixth International Symposium on Temperature: Its Measurement and Control in Industry,* Washington, D.C., Princeton Plasma Physics Laboratory Report, PPPL-1891.

Bretz, N., and G. Taylor (1982). TFTR Physics Group Report No. **47.**

Brower, D. L. (1984). *Characterization of Microturbulence in TEXT by Multichannel Scattering of Far-Infrared Laser Radiation,* Ph.D. dissertation, UCLA.

Brower, D. L., et al. (1985a). *Proc. of 10th Int. Conf. on Plasma Physics and Controlled Nuclear Fusion Research,* IAEA-CN-44/A-IV-6, **1,** 273.

Brower, D. L., W. A. Peebles, N. C. Luhmann, Jr. (1985c). UCLA Center for Plasma Physics and Fusion Engineering, Rep. No. PPG-862.

Brower, D. L., C. P. Ritz, W. A. Peebles, N. C. Luhmann, Jr., E. J. Powers (1986). *Int. J. of Infrared and Millimeter Waves.* **7,** 447–458.

Chen, Chao-Chun (1973). *IEEE Trans. Microwave Theory Tech.* **MTT-21,** no. 1, 1.

Chen, H. H., and C. S. Liu (1977). Phys. Rev. Lett., **39,** 1147.

Chen, L., and C. Z. Cheng (1980). *Phys. Fluids* **23,** 2242.

Cheo, P. K., private communications; also see Rubino, R. A., Farina, J. D. and Cheo, P. K. (1985). *Bull. Am. Phys. Soc.* **30,** 1586.

Clayton, C. E., H. E. Huey, C. Pawley, A. Y. Lee, F. F. Chen, C. Joshi, N. C. Luhmann, Jr. (1982). *Proc. 9th Int. Conf. on Plasma Physics and Controlled Nuclear Fusion Research,* IAEA-CN-41/B-1-2, Baltimore, MD.

Cohen, B. I., A. N. Kaufman, K. M. Watson (1972). *Phys. Rev. Lett.,* **29,** 581.

Cohen, B. I., M. A. Mostrom, D. R. Nicholson, A. N. Kaufman, A. B. Langdon (1975). *Phys. Fluids,* **18,** 470.

Cohen, B. I. (1975). Ph.D. dissertation, University of California.

Cohen, B. I. (1984). *Comments on Plasma Physics and Controlled Fusion* **8,** 197.
Crenn, J. P., and D. Veron (1974). *Proc. of 2nd Int. Conf. on Infrared Physics,* Zurich, Switzerland, March 1979.
Crenn, J. P. (1979a). *A Study of Guided Propagation for Infrared Interferometry In Tokamaks.* Report EUR-CEA-FC-924.
Crenn, J. P. (1979b). *IEEE Trans. Microwave Theory Tech.* **MTT-26,** no. 6, 573.
Crenn, J. P., and D. Veron, (1981). Wave guiding of far infrared radiation for the JET interferometer, *Proc. of 6th Int. Conf. on Infrared and Millimeter Waves,* IEEE cat. no. 81 CH1645-1 MTT, 4-10.
Crowley, J. D. (1983). SPIE Proc., **423,** 30.
Danielewicz, E. J. (1980). *Submillimeter Local Oscillators for Heterodyne Spectroscopy,* Aerospace Corp. Report ATR-80 (7696)-1.
Davis, S. L., S. S. Medley, M. Brusati (1979). Princeton Plasma Physics Laboratory Report PPPL-1478.
Degnan, J. J. (1976). *Appl. Phys.,* **11,** 1.
DeTemple, T. A., E. J. Danielewicz (1983). Continuous-wave optically pumped lasers, chapter in *Coherent Sources and Applications,* Part II of *Infrared and Millimeter Waves,* **7,** edited by K. J. Button, Academic Press, New York.
DiVergilio, W. F., A. Y. Wong, H. C. Kim, Y. C. Lee (1977). *Phys. Rev. Lett.,* **38,** 541.
Dodel, G., and W. Kuntz (1978). *Infr. Phys.,* **18,** 773.
Dodel, G., H. Hailer, E. Holzhauer, P. N. Ignacz, J. H. Massig, H. P. Roser, H. Salzmann, F. Soldner, T. Vogel (1984), *IR and MM Waves Conf.* in Takarazuka, Japan.
Doyle, E. S., D. E. Evans, D. Frigone, M. von Hellerman, A. Murdoch (1983). *Plasma Phys.,* **25.**
Elsasser, H., and H. Schamel (1977). *Plasma Phys.,* **19,** 1055.
Erickson, N. R. and H. R. Fetterman (1982). *Proc. of Fourth APS Top. Conf. on High Temperature Plasma Diagnostics,* Boston, MA.
Evans, D. E., M. von Hellerman, E. Holzhauer (1982). *Plasma Phys.,* **24,** 819.
Fetterman, H., P. E. Tannenwald, B. J. Clifton, C. D. Parker, W. D. Fitzgerald, and N. R. Erickson (1978). *Appl. Phys. Lett.,* **33,** 151.
Fetterman, H. R., B. J. Clifton, C. D. Parker, and P. E. Tannenwald (1980). *Proceedings of the Japan-USA Workshop on Far-Infrared Diagnostics,* Cambridge, MA.
Fetterman, H. R., T. Sollner, P. Parrish, C. Parker, P. E. Tannenwald (1980). *Proceedings of the Japan-USA Workshop on Far-Infrared Diagnostics,* Cambridge, MA.
Fetterman, H. R. and W. Chew (1985). private communications.
Foote, F. B., and D. T. Hodges (1979). Proc. 4th Int. Conf. on Infrared and Millimeter Waves, Florida, IEEE cat. no. CH 1384-7 MIT, 78.
Gordon, E.I. (1966). *Proc. IEEE,* **54,** 1391.
Gustincic, J. J. (1977a). *Proc. Soc. of Photo-Opt. Instrum. Eng.,* **105,** 40.
Gustincic, J. J. (1977b). *IEEE MTT-S International Microwave Symposium Digest,* IEEE 77 CH 1219-5 MTT, New York.
Gustincic, J. J., Th. de Grauuw, D. T., Hodges, and N. C. Luhmann, Jr. (1977), post deadline paper at *IEEE MTT-S International Microwave Symposium.*
Hasegawa, A., and Y. Kodoma (1978). *Phys. Rev. Lett.* **41,** 1470.
Hirschfield, J. L. (1980), in *Infrared and Millimeter Waves,* **1,** edited by K. J. Button, Academic Press, New York.
Hodges, D. T., F. B. Foote, R. D. Reel (1976). *Appl. Phys. Lett.* **29,** 662.
Horton, W., and R. D. Estes (1979). *Nucl. Fusion* **19,** 203.
Horton, W. (1984). in *Handbook of Plasma Physics,* **2,** North Holland Publishing Co., Amsterdam, 383.

Huey, H. E., A. Mase, N. C. Luhmann, Jr., W. F. DiVergilio, J. J. Thomson (1980). *Phys. Rev. Lett.,* **45,** 795.

Huey, H. E., N. C. Luhmann, Jr., C. J. Pawley (1984). in *Laser Interaction and Related Plasma Phenomena,* **6,** edited by H. Hora and G. H. Miley, Plenum Publishing Co., 605.

Hugenholtz, C. A. J., and B. J. H. Meddens (1979). *Rev. Sci. Instrum.,* **50,** 1123.

Hugenholtz, C. A. J., and B. J. H. Meddens (1982). *Rev. Sci. Instrum.,* **53,** 171.

Hunt, G. F., R. C. Kaifer, D. R. Slaughter, R. E. Strout, D. W. Rueppel (1980). *IEEE Trans. Nucl. Sci.* **NS-27,** 757.

Hwang, D. Q., J. Hosea, H. Thompson, J. R. Wilson, S. Davis, D. Herndon, R. Kaita, D. Mueller, S. Suckewer, C. Daughney, P. Colestock (1983). *Phys. Rev. Lett.,* **51,** 1865.

Ishibashi, T., M. Ino, T. Makimura, M. Ohmari (1979). *Electron Lett.,* **13,** 299.

Jacobsen, A. R. (1978). *Rev. Sci. Instrum.,* **49,** 673.

Jacobsen, A. R., and L. J. Jolin (1981). *Proc. of SPIE,* **288,** 269.

James, C. R., and W. B. Thomson (1987). *Can. J. Phys.,* **45,** 1771.

Jones, W. D., L. Z. Kennedy, J. W. Bilbra, H. B. Jeffreys (1984). *Appl. Opt.,* **23,** 730.

Joyce, G. R., D. W. Choi, R. D. Bengtson, E. J. Powers, D. L. Brower, The application of digital complex demodulation to the reduction of interferometer data, submitted to *Rev. Sci. Instrum.*

Kaita, R., and S. S. Medley (1979). Princeton Plasma Physics Laboratory Report PPPL-1582.

Kantorowicz, G., and P. Palluel (1979) in *Infrared and Millimeter Waves,* **1,** ed. by K. J. Button, Academic Press, New York.

Kasparek, W., K. Hirsch, E. Holzhauer (1980). *Plasma Phys.,* **22,** 555.

Kasparek, W., and E. Holzhauer (1981). Institut für Plasmaforschung der Universitat Stuttgart, report No. IPF-81-5.

Kim, H. C., R. L. Stenzel, A. Y. Wong (1974). *Phys. Rev. Lett.,* **33,** 886.

Kneubuhl, F. K., and Ch. Sturzenegger (1980) in *Infrared and Millimeter Waves,* **3,** ed. by K. J. Button, Academic Press, New York.

Krautle, H., E. Sauter, G. V. Schultz (1977). *Infrared Phys.,* **17,** 437.

Krautle, H., E. Sauter, G. V. Schultz (1978). *Infrared Phys.,* **18,** 705.

Kristal, R. (1978). in *Diagnostics for Fusion Experiments,* ed. by E. Sindoni and C. Wharton, Permagon Press, Oxford.

Kroll, N., A. Ron, N. Rostoker (1984). *Phys. Rev. Lett.,* **13,** 83.

Kuno, H. J. (1981). Solid state millimeter wave power sources and combiners, *Microwave Journal,* **24,** 21.

Lee, A. Y., Y. Nishida, N. C. Luhmann, Jr., S. P. Obenschain, B. Gu, M. Rhodes, J. R. Albritton, E. A. Williams (1982). *Phys. Rev. Lett.,* **48,** 319.

Lee, P., R. J. Taylor, W. A. Peebles, H. Park, C. X. Yu, Y. Xu, N. C. Luhmann, Jr., S. X. Jin (1982a). *Phys. Rev. Lett.,* **49,** 205.

Lee, P., N. C. Luhmann, Jr., H. Park, W. A. Peebles, R. J. Taylor, Y. Xu, C. X. Yu (1982b). *Proc. of USA-Japan Workshop on Submillimeter Diagnostic Techniques,* Nagoya, Japan, US-J FRC-006.

Lehecka, T., Q. X. Zu, M. Du, R. Dworak, R. L. Savage, Jr., W. A. Peebles, N. C. Luhmann, Jr. (1985). *Rev. Sci. Instrum.,* **56,** 940.

Lipschultz, B., B. LaBombard, E. S. Marmar, M. M. Pickrell, J. L. Terry, R. Watterson, S. M. Wolfe (1984). *Nucl. Fusion,* **24,** 977.

Lovelace, R. V. E., E. E. Salpeter, L. E. Sharp, D. Harris (1970). *Am. Astron. J.,* **159,** 1047.

Luhmann, N. C., Jr. (1979). in *Infrared and Millimeter Waves,* **2,** edited by K. J. Button, Academic Press, New York.

Luhmann, N. C., Jr., and W. A. Peebles (1985), *Rev. Sci. Instr.,* **55,** 279.

Luhmann, N. C., Jr., and W. A. Peebles, "Laser diagnostics of Magnetically Confined Thermonuclear Plasmas" Chapter 5, volume 5, pg. 455, *Laser Handbook* ed. by M. Bass and M. L. Stitch, North Holland Publishing Co., Amsterdam.

Ma, C. H., D. P. Hutchinson, P. A. Staats, K. L. Vander Sluis (1982). *Int. Jour. of Infrared and Millimeter Waves,* **3,** 263.

Mansfield, D. K., L. C. Johnson, A. Mendelsohn (1980). *Int. Jour. of Infrared and Millimeter Waves,* **1,** 631.

Marcateli, E. A., and R. A. Schmeltzer (1964). *Bell Sys. Tech. J.,* **43,** 1783.

Mase, A., N. C. Luhmann, Jr., J. Holt, H. Huey, M. Rhodes, W. F. DiVergilio, J. J. Thomson, C. J. Randall (1981). *Proc. of 8th IAEA Conf. on Plasma Physics and Controlled Nuclear Fusion Research,* **II.,** 745.

Medley, S. S. (1981). Princeton Plasma Physics Laboratory Report, PPPL-1770.

Medley, S. S., and R. Persing (1981). *Rev. Sci. Instr.,* **52,** 1463.

Mizuno, K., M. Ohmoni, K. Miyazawa, M. Morimoto, S. Kodaira, and S. Ono (1979). *Infrared Phys.,* **13,** 401.

Mohamed-Benkadda, M. S., J. L. Gerstenmayer, D. Gresillon, A. Hemon, D. Bense, T. Viala (1984). Three wave coupling in a fluid turbulence experimental bispectrum, *Int. Conf. on Plasma Physics,* **II,** 311, Lausanne, Switzerland.

Morales, G. J., and Y. C. Lee (1977). *Phys. Fluids,* **20,** 1135.

Nexsen, W. E., Jr., W. C. Turner, W. F. Cummins (1979). *Rev. Sci. Instr.,* **50,** 1227.

Obenschain, S. P., N. C. Luhmann, Jr. (1977). *Appl. Phys. Lett.,* **30,** 452.

Obenschain, S. P., N. C. Luhmann, Jr. (1979). *Phys. Rev. Lett.,* **42,** 311.

Palluel, P., and A. K. Goldberger (1956). *Proc. IRE,* **44,** 333.

Pan, D. S., and N. Lee (1981). IEEE Rep. no. 81, CH 1645-1MTT.

Park, H., W. A. Peebles, A. Mase, N. C. Luhmann, Jr., A. Semet (1980). *Appl. Phys. Lett.,* **37,** 279.

Park, H. K. (1980). *Characteristics of a Far-Infrared Laser Scattering Apparatus for Plasma Wave Studies',* Masters thesis, UCLA.

Park, H., C. X. Yu, W. A. Peebles, N. C. Luhmann, Jr., R. Savage (1982). *Rev. Sci. Instr.,* **53,** 1535.

Park, H., N. C. Luhmann, Jr., W. A. Peebles, R. Kirkwood (1984a). *Phys. Rev. Lett.,* **52,** 1609.

Park, H., P. S. Lee, W. A. Peebles, N. C. Luhmann, Jr. (1984b). UCLA Center for Plasma Physics and Fusion Engineering, Rep. No. PPG-788.

Park, H. K. (1984). *Tokamak Plasma Waves Studies via Multi-Channel Far-Infrared Laser Scattering,* Ph.D. dissertation, UCLA, 1984.

Park, H., D. L. Brower, W. A. Peebles, N. C. Luhmann, Jr., R. L. Savage, Jr., C. X. Yu, (1985). *Rev. Sci. Instrum.,* **56,** May/June.

Parrish, P. T., T. C. L. G. Sollner, R. H. Matthews, H. R. Fetterman, C. D. Parker, P. E. Tannenwald, A. G. Cardiasmenos (1982). Printed dipole-Schottky diode mixer array, *Proc. SPIE Symposium* **337.**

Pawley, C. J., H. E. Huey, N. C. Luhmann, Jr. (1982). *Phys. Rev. Lett.* **49,** 877.

Pawley, C. J. (1985). private communication.

Peebles, W. A., and M. J. Herbst (1978). *IEEE Trans. Plasma Sci.,* **PS-6,** 564.

Peebles, W. A., and T. Lehecka et al., (1985). *Rev. Sci. Instrum.* **56,** 940.

Potter, P. D. (1973). JPL Technical Report 32-1526, **XIX,** 55–62, Jet Propulsion Laboratory, Pasadena, CA.

Raven, A., and D. T. Rumsby (1977). *Phys. Lett.,* **60A,** 42.

Rhodes, M., A. Y. Lee, Y. Nishida, N. C. Luhmann, Jr., S. P. Obenschain (1979). UCLA Center for Plasma Physics and Fusion Engineering Report, PPG-428 (1979).

Ritz, C. P., and E. J. Powers (1985). *Estimation of Nonlinear Transfer Functions for Fully Developed Turbulence,* University of Texas at Austin, Fusion Research Center Report, FRCR 273.

Rosenbluth, M. N., and C. S. Liu (1972). *Phys. Rev. Lett.,* **29,** 701.

Rosenbluth, M. N., and P. H. Rutherford (1981). in *Fusion,* ed. by E. Teller, Academic Press, New York, 31.

Saito, T., Y. Hamada, T. Yamashita, M. Ikeda, S. Tanuka (1981). *J. Appl. Phys.,* **52,** 5305.

Schamel, H., and K. Elsasser (1978). *Plasma Phys.,* **20,** 837.

Schmidt, G. (1973). *Phys. Fluids,* **16,** 1676.

Schmitt, J. P. M. (1973). *Phys. Rev. Lett.,* **31,** 982.

Schmitt, J. P. M., P. Krumm (1976). *Phys. Rev. Lett.,* **37,** 753.

Semet, A., A. Mase, W. A. Peebles, N. C. Luhmann, Jr., S. Zweben (1980). *Phys. Rev. Lett.,* **45,** 454.

Sheffield, J., *Plasma Scattering of Electromagnetic Radiation,* Academic Press, 1975.

Similon, P. L., P. H. Diamond (1984). *Phys. Fluids,* **27,** 916.

Sindoni, E., and C. Wharton, editors (1978). *Diagnostics for Fusion Experiments,* Pergamon Press, Oxford.

Slaughter, D. R., and W. L. Pickles (1979). *Nucl. Inst. and Methods,* **160,** 87.

Slaughter, D. R. (1982). *APS Topical Conference on High Temperature Plasma Diagnostics,* Boston, MA.

Slusher, R. E., and C. M. Surko (1978). *Phys. Rev. Lett.,* **40,** 400.

Slusher, R. E., and C. M. Surko (1980). *Phys. Fluids,* **23,** 472.

Slusher, R. E., C. M. Surko, J. F. Valley, T. Crowley, E. Mazzucato, K. McGuire (1984). *Phys. Rev. Lett.,* **53,** 667.

Smith, D. E., E. J. Powers, G. S. Caldwell (1974). *IEEE Transactions on Plasma Sci.,* **PS-2,** 263.

Soltwisch, H. (1983). Measurements of electron and current density profiles in TEXTOR by means of a far-infrared laser interferometer/polarimeter, *11th European Conference on Controlled Fusion and Plasma Physics,* **7D,** Part 1, 123–127.

Sonoda, Y., A. Sakurai, K. Muraoka, M. Akasahi (1982). *Jpn. J. Appl. Phys.,* **21,** 32.

Strachen, J. P., P. Colestock, H. Eubank, L. Grisham, J. Hovey, G. Schilling, L. Stewards, W. Stodiek, R. Stooksberry, K. M. Young (1979). *Nature,* **279,** 626.

Suchewer, S., and E. Hinnov (1978). *Phys. Rev. Lett.,* **41,** 756.

Surko, C. M., and R. E. Slusher (1983). *Science* **221,** 817.

Symons, R. S., and J. R. Jory (1981). *Adv. Electron. Electron Phys.,* **55,** 1.

Taylor, R. J., P. Lee, N. C. Luhmann, Jr., A. Mase, G. J. Morales, W. A. Peebles, A. Semet, F. Schwirghe, S. Talmadge, S. Zweben, M. A. Hedemann, B. S. Levine, R. W. Gould (1980). *Plasma Physics and Controlled Nuclear Fusion Research,* **2,** 61.

TFR Group (1978). Tokamak plasma diagnostics, *Nucl. Fusion,* **18,** 5.

TFR Group, A. Truc, D. Gressillon (1982). *Nucl. Fusion,* **22,** 1577.

Thomson, C. S. F., (1981). Rapport Technique no. 2451 T.

Tsukishima, T., and O. Asada (1978). *Japan J. Appl. Phys.,* **17,** 2059.

Tsukushima, T. (1979) in *Diagnostics for Fusion Experiments* edited by E. Sindoni and C. Wharton, Permagon Press, Oxford, 255.

Tsukishima, T., O. Asada, K. Yoshioka, and A. Inoue (1980). Measurements of nonstationary and anisotropic density fluctuations using electromagnetic wave scattering, *US-Japan FIR Workshop,* Boston, MA.

Tsukishima, T., M. Tago, I. Nishida, H. Inuzuka, M. Nagatsu, and K. Mizuno (1984). A new homodyne spectroscopy in far-infrared regime, *Int. Conf. on Infrared and Millimeter Waves,* **M-5-2,** 49, Takarazuka, Japan.

Veron, D. (1974). *Optics Comm.,* **10,** 95.

Veron, D., P. Belland, M. J. Beccaria (1978). *Infrared Physics,* **18,** 465.

Veron, D. (1979) in *Infrared and Millimeter Waves,* 2, edited by K. J. Button, Academic Press, New York.

Veron, D. (1980). Line electron density measurement in the JET plasma, *IAEA Workshop on Plasma Diagnostics,* Princeton, N. J.

Veron, D. (1980). *Proc. of Int. School on Plasma Physics, Varenna.*

Vogel, T., G. Dodel (1984). *Acousto-Optic Modulation in the Far-Infrared,* Institut für Plasmaforschung, Universitat Stuttgart, 7000 Stuttgart 80, Federal Republic of Germany. Technical Report.

Waltz, R. E., W. Pfieffer, R. R. Dominguez (1980). *Nucl. Fusion,* **20,** 43.

Weibel, E. S. (1976a). *Phys. Fluids,* **19,** 1237.

Weibel, E. S. (1976b). *Phys. Rev. Lett.,* **37,** 1619.

Wolfe, S. M., K. J. Button, J. Waldman, D. R. Cohn (1976). *Appl. Optics,* **17,** 2645.

Wong, A. Y. (1976) in *Laser Interaction and Related Plasma Phenomena,* **4B,** edited by H. J. Schwartz and H. Hora, Plenum Press, New York.

Woskoboinikow, P., D. R. Cohn and R.J. Temkin (1982). MIT Report no. PFC/JA-82-9.

Woskoboinikow, P., W. J. Mulligan, N. J. Erickson, D. R. Cohn, R. J. Temkin, H. R. Fetterman, B. Lax (1983). *Proc. 7th Int. Conf. on Infrared and Millimeter Waves,* Marseille, France.

Wurden, G. A., M. Ono, K. L. Wong (1982). *Phys. Rev.,* **A26,** 2297.

Wurden, G. A. (1982). Ph.D. dissertation, Princeton University.

Yamanaka, M. (1977). *J. Opt. Soc. Am.,* **76,** 959.

Yamanaka, M., Y. Takeda, S. Tanigawa, N. Nishizawa, N. Noda, J. Fujita, M. Takai, M. Shimobayashi, Y. Hayashi, T. Koizumi, K. Nagasaka, S. Okajima, Y. Tsunawaki, A. Nagashima (1980). *Int. Jour. of Infrared and Millimeter Waves,* **1,** 57.

Young, P. E. (1984). *Multichannel Far-Infrared Interferometer/Polarimeter,* Ph.D. dissertation, UCLA.

Yu, C. X., B. W. James, W. A. Peebles, N. C. Luhmann, Jr. (1983). *Proc. 7th Int. Conf. on Infrared and Millimeter Waves,* Marseille, France.

Zah, C., Lam, W., Smith, J. S., Rav-Noy, Z., and Rutledge, D. B. (1984). *Proc. of 9th Int. Conf. on Infrared and Millimeter Waves,* 248, Takarazuka, Japan.

CHAPTER 4

Special Imperfections in Semiconductors

E. Otsuka

Department of Physics
Osaka University
Toyonaka, Osaka 560, Japan

I. Introduction

Historically, the study of crystalline defects or imperfections in semiconductors has mostly been treated by DC transport or by optical methods. In conjunction with metallurgy, microscopic observation of dislocation loops or stacking faults also helped in visualizing certain aspects of imperfections. Relatively few approaches are known, however, from the infrared and millimeter wave side, except possibly definite impurity spectroscopy or scattering.

It should be strongly emphasized that defects other than simple impurities, or complexes adhering to uncommon impurities, can also be tractable within the scope of this book.

The present author is of the opinion that the magneto-optical method

ISBN 0-12-147700-2

making use of a coherent radiation source is, as elsewhere, immensely helpful in disclosing various hidden features of an obscurely known defect. Unfortunately, the number of relevant research workers is relatively small. Therefore the method is neither widely utilized, nor approved as a general breakthrough. A partial reason for the hesitation on the part of defect-studying people to use millimeter waves or far-infrared is based on the belief that the technology needed is difficult in comparison with ordinary DC transport or photoluminescence measurements. Such a belief is almost superstitious, but, in the experience of the author, very hard to destroy. It is also true, on the other hand, that once the misleading belief is overcome, an experimenter can enjoy a free new world. Fortunately, the readers of this book are expected to have little fear of technology in long-wave optics. It is also true, on the other hand, that high technologists are frequently unaware of physical topics. It is the purpose of this chapter to introduce some items of physical topics on special defects in semiconductors, successfully treated or being treated by the use of long wavelength magneto-optics, encouraging the technological specialists to make more contribution to an important branch of semiconductor physics.

II. Electron Scattering by Dislocations

A. Electron Cyclotron Resonance in Plastically Deformed Germanium

Crystal dislocations have seldom been treated with radio wave techniques. For germanium, in particular, perhaps the sole piece of work was made by the author (Otsuka and Yamaguchi, 1967). It will be reproduced here to some extent and discussed in conjunction with related topics.

An ultrapure ingot of germanium, slightly p-type and having a room temperature resistivity of 70 ohm-cm, was sliced into rectangular plates, typically to a size of $1.0 \times 1.0 \times 1.5$ mm^3. After being etched by CP4 and cleaned with KCN, each plate was bent to a desired degree, with the help of stainless steel edges (Fig. 1) in an atmosphere of nitrogen gas at 500°C. Then the bent sample was annealed at 650°C for 24 hours, being flashed by oxygen gas that passed through a distilled water vessel. Of all the possible types of dislocation, only 60° and 90° dislocations were studied. The two angles are those between the Burgers vector and the dislocation axis. The former is born as one bends the crystal about a $\langle 110 \rangle$ axis, while the latter about a $\langle 111 \rangle$ axis. Geometry of the dislocation axis, bending axis, resonant microwaves and magnetic field is described in Fig. 2. Microwaves and the magnetic field are two essentials in cyclotron resonance. 30 GHz direct detection or a 35 GHz superheterodyne detection system was em-

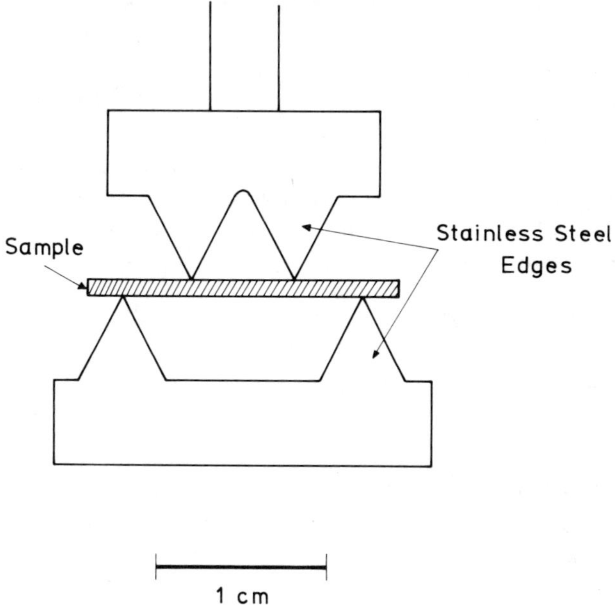

FIG. 1 Side view of the stainless steel edges used to bend germanium crystals (Otsuka and Yamaguchi, 1967).

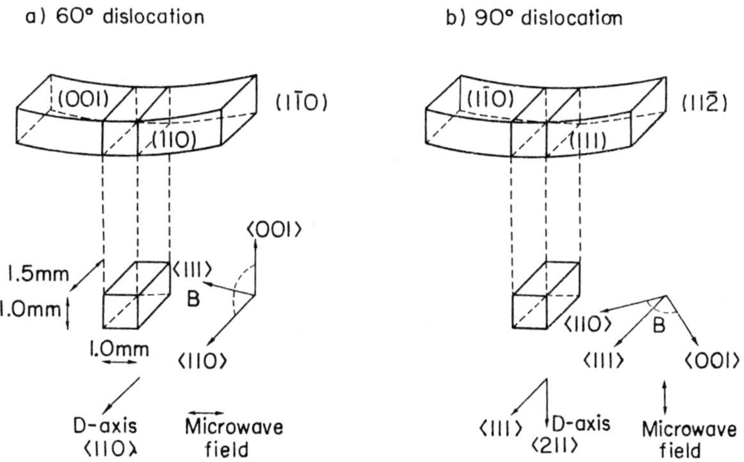

FIG. 2 Ways of cutting the bent crystal and rotating the external magnetic field for both (a) 60° and (b) 90° dislocations. The magnetic field is rotated in the (110) plane for the microwave cyclotron resonance experiment. The direction of the dislocation line is indicated by the "*D*-axis" arrow (Otsuka and Yamaguchi, 1967).

ployed, depending on the purpose of measurement. Temperature was varied between 1.5 and 4.2 K. Carriers were generated by a tungsten lamp at a light-chopping frequency of 450 Hz.

Two things can be measured in cyclotron resonance. One is the linewidth and the other is the intensity of the absorption signal. The former gives the scattering rate while the latter gives the carrier density, and hence the relative lifetime of the photo-excited carriers. What we see is the resonance of electron having the lightest cyclotron mass when the magnetic field is applied along the direction of $\langle 111 \rangle$. In general, the linewidth consists of the contributions from various sources; namely, lattice scatterings, carrier–carrier scatterings, carrier–impurity scatterings and, in the present special case, carrier–dislocation scatterings. These contributions are assumed to satisfy the relation:

$$\frac{1}{\tau} = \frac{1}{\tau_L} + \frac{1}{\tau_{\text{cc}}} + \frac{1}{\tau_I} + \frac{1}{\tau_d}, \tag{1}$$

where τ_L, τ_{cc}, τ_I, and τ_d are the electron scattering times by lattice, carriers, impurities and dislocations, respectively. τ is the total scattering time that satisfies the formula (Otsuka et al., 1966; Otsuka, 1980)

$$\omega_c \tau = \frac{2B_r}{\Delta B}, \tag{2}$$

where ω_c is the cyclotron frequency, or $\omega_c = eB_r/m^*$, B_r is the resonance field and ΔB is the full width of the absorption line that is assumed to be a Lorentzian. The relation in Eq. (2) is only an approximation in defining the scattering relaxation time τ, but this definition turns out to be justifiable for most practical purposes. So is the relation in Eq. (1). As for the second and third terms on the right-hand side of Eq. (1), one can exclude these by taking an extremely weak illumination and an extremely pure sample. In our case, such conditions are fulfilled, so that all the contribution is coming from scatterings by lattice vibrations and dislocations. The lattice scattering is dominantly caused by acoustical phonons at liquid helium temperatures. For the relevant electron resonance in Ge, obtained in the geometry of $B \parallel \langle 111 \rangle$, one has secured the relation

$$\frac{1}{\tau_L} = aT^{3/2} \tag{3}$$

with

$$a = 4.8 \times 10^8 \text{s}^{-1}\text{K}^{-3/2} \tag{3a}$$

for the frequency range of 30–35 GHz (Otsuka et al., 1966; Ohyama et al., 1970). It is accordingly easy to subtract the contribution from the lattice scattering. The remaining portion after this subtraction gives the genuine contribution of $1/\tau_d$. Of course $1/\tau_d$ is expected to depend on the density of dislocations. We have prepared densities of $N_d = 5 \times 10^6$ cm^{-2}, 1.0 × 10^7 cm^2 and 1.2 × 10^7 cm^{-2} for 60° dislocations, and similarly $N_d = 5 \times 10^6$ cm^{-2} and 1.1 × 10^7 cm^{-2} for 90° dislocations. These figures are obtained after averaging the results of etch-pit countings at various areas of the etched surface in microscopic observation.

In Fig. 3, the temperature dependence of $1/\tau$ is shown both for 60° and 90° dislocations, with two densities each. Solid straight lines are the contribution from lattice scatterings as given by Eq. (3). Genuine contribution from dislocations is always found to be practically independent of temperature between 1.5 and 4.2 K.

Let us examine next the dependence of $1/\tau_d$ on densities of dislocations. Values of derived $1/\tau_d$ is plotted in Fig. 4 against N_d for two types of dislocations. Though there exists a considerable scatter in the estimate of N_d, we may draw out a dependence of $N_d^{1/2}$. One more feature is the difference in magnitude: 90° dislocations scatter electrons stronger than 60° dislocations.

Finally, temperature dependence of the signal intensity is shown in Fig. 5 for a particular sample, having the density of 5.0 × 10^8 cm^{-2} of 90° dislocations. This is a rather crude analysis, since it is very difficult to hold the sensitivity of the equipment unchanged in variance of the bath temperature in which the sample and the nonresonant microwave waveguide are immersed. Nevertheless, our repeated experience clearly indicates more enhanced signal at lower temperatures. The dashed line, giv-

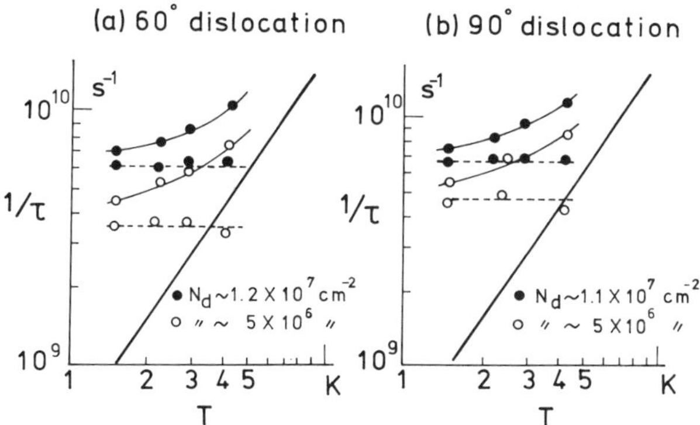

FIG. 3 The inverse relaxation time of electrons versus temperature for both 60° and 90° dislocations. The solid straight line in each figure indicates the linewidth due to the lattice scattering, i.e., the linewidth for an ultrapure sample. Circles along the dashed lines are after subtraction of the contribution from the lattice scattering (Otsuka and Yamaguchi, 1967).

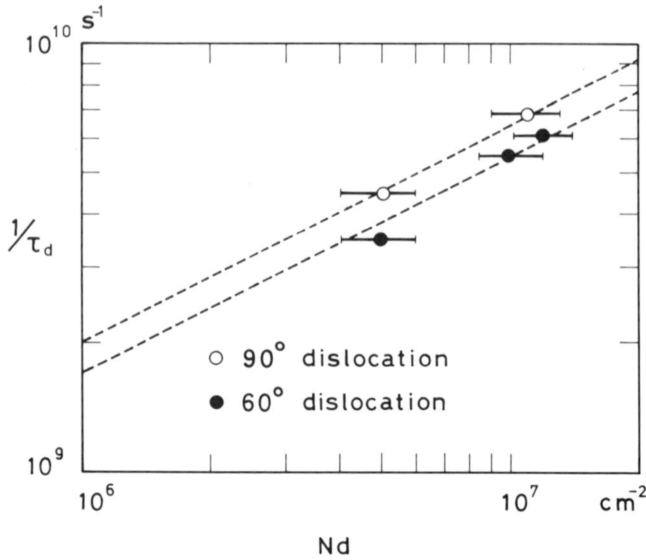

FIG. 4 Inverse collision time of electron with dislocations is plotted against the concentrations of dislocations for both 60° and 90° dislocations. The dashed lines are to indicate $N_d^{1/2}$ dependence (Otsuka and Yamaguchi, 1967).

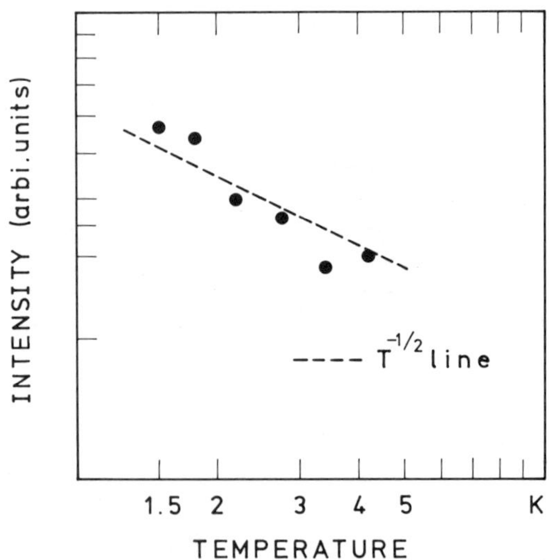

FIG. 5 Intensity of electron cyclotron resonance versus temperature for a particular Ge sample having 5×10^6 cm^{-2} of 90° dislocations. The dashed line is to indicate $T^{-1/2}$ dependence (Otsuka and Yamguchi, 1967).

ing a $T^{-1/2}$ dependence, is drawn not to fit the scattered experimental data, but to indicate a consistency with a theory described later.

B. Theoretical Treatment of Cyclotron Resonance Linewidth

1. *An Orthodox Transport Theory*

Effects of dislocations on mobilities in semiconductors were treated, almost once and for all, by Dexter and Seitz (1952). Their idea is essentially to take the scattering potential proportional to the deformation potential caused by the dilatation around an edge-type dislocation; in other words, to put the scattering potential $V(r)$ as

$$V(r) = \frac{E_1}{e}\,\Delta(r) \tag{4}$$

where $\Delta(r)$ is the dilatation of the lattice at the point r, and E_1 is the so-called deformation potential constant, giving the shift of the band edge at a unit strain. The amount of the dilatation around an edge-type dislocation has been derived in various textbooks (see, e.g., Cottrell, 1953) on dislocations to yield

$$\Delta(r) = -\frac{b}{2\pi}\,\frac{1-2\nu}{1-\nu}\,\frac{\sin\theta}{r}. \tag{5}$$

Here b is the magnitude of the Burgers vector for a standard edge-type dislocation, ν the Poisson ratio, r the distance from the dislocation line and θ the angle from the slip direction. After carrying out a necessary calculation of the relevant matrix elements for the electron scattering between wave vector states k and k' (Dexter, 1952), one arrives at an expression

$$\frac{1}{\tau_d} = \frac{3\pi}{32}\,\frac{E_1^2 b^2}{k_B T h}\left(\frac{1-2\nu}{1-\nu}\right)^2 N_d. \tag{6}$$

This is the theoretical prediction by Dexter and Seitz (1952). Its form is already out of agreement with our cyclotron resonance result that shows $1/\tau_d \propto N_d^{1/2}T^0$. Moreover, the magnitude predicted by Eq. (6) is considerably larger than cyclotron resonance observation. In fact, after putting the numerical values of E_1 etc. into Eq. (6) (Murase et al., 1970), one obtains

$$\frac{1}{\tau_d} = 6.6\times 10^4 N_d T^{-1}\mathrm{s}^{-1} \tag{7}$$

that is to compare with

$$\frac{1}{\tau_d} = 2.0 \times 10^6 N_d^{1/2} s^{-1} \tag{8a}$$

for 90° dislocations and

$$\frac{1}{\tau_d} = 1.7 \times 10^6 \, N_d^{1/2} \, s^{-1} \tag{8b}$$

for 60° dislocations. Here N_d is counted in cm^{-2} as before.

2. *Cyclotron Resonance Approach*

To account for the discrepancy between the cyclotron resonance result and the above transport theoretical prediction, one is urged to derive an independent treatment more appropriate in explaining the line-broadening of cyclotron resonance. We saw before a similar disagreement between the transport theoretical prediction and the cyclotron resonance for ionized impurity scatterings (Otsuka et al., 1973; Matsuda and Otsuka, 1979; Otsuka, 1980). Such a disagreement occurs when the scattering potential for electrons is of a long-range nature, specifically if the force range of the scattering potential is larger than the cyclotron radius. In fact, both ionized centers and dislocations yield scattering potentials proportional to $1/r$, with only an angular factor of $\sin \theta$ incorporated in the dislocation potential. Of course a dimensional difference should also be taken into account. In the dislocation scattering, the radial magnitude r bears a two-dimensional nature, while in the ionized impurity or carrier–carrier scattering the same is responsible for three-dimensional distance. Nevertheless, many of the arguments made in the ionized impurity or carrier–carrier scattering are valid also in the dislocation scattering. The most important common feature is that the frequency shift in cyclotron motion caused by the long-range scattering potential is given by

$$\Delta\omega = \left(\frac{1}{2} m^* \omega_c\right) (V_{xx} + V_{yy}) \tag{9}$$

where V_{xx} and V_{yy} are the electric field gradients or the second derivatives of the scalar potential, the z-axis being taken along the magnetic field (Kawamura et al., 1964). In the dislocation scattering case, one finds

$$V_{xx} + V_{yy} = 2 \frac{E_1 b}{2\pi} \frac{1 - 2\nu}{1 - \nu} \sin\theta \frac{3 - 4\sin^2\theta}{r^3} \tag{10a}$$

for $B \parallel x$,

$$V_{xx} + V_{yy} = 2 \frac{E_1 b}{2\pi} \frac{1 - 2\nu}{1 - \nu} \sin\theta \frac{4\sin^2\theta - 3}{r^3} \tag{10b}$$

for $B \parallel y$,

and

$$V_{xx} + V_{yy} = 0 \qquad \text{for} \quad B \parallel z. \tag{10c}$$

We shall take the mean square average of these three representative geometries. Then we obtain

$$\langle \Delta\omega^2 \rangle = \frac{4}{3}\left[\frac{1}{2m^*\omega_c}\frac{E_1 b}{2\pi}\left(\frac{1-2\nu}{1-\nu}\right)\right]^2$$

$$\times \int_0^{2\pi}\int_{R_c}^{(\pi N_d)^{-1/2}} \frac{4\sin^2\theta\,(3-4\sin^2\theta)^2}{r^5}\,\frac{dr d\theta}{\int_0^{2\pi}\int_0^{(\pi N_d)^{-1/2}} r dr d\theta}$$

$$= \left[\frac{E_1 b}{2\pi m^*\omega_c}\left(\frac{1-2\nu}{1-\nu}\right)\right]^2 \frac{\pi N_d}{3R_c^4}. \tag{11}$$

We introduce here the cutoff radius R_c around the dislocation axis. This is to prevent the integral from diverging. The physical meaning of R_c will be discussed later. If we replace $\langle\Delta\omega^2\rangle^{1/2}$ by $1/\tau_d$, we have

$$\frac{1}{\tau_d} = \left(\frac{N_d}{3\pi}\right)^{1/2}\left(\frac{|E_1|b}{m^*\omega_c}\right)\frac{\left(\frac{1-2\nu}{1-\nu}\right)}{2R_c^2}. \tag{12}$$

This form essentially accounts for the general behavior of the cyclotron resonance. Indeed, Eq. (12) contains a proportionality to $N_d^{1/2}$ and no temperature dependence. Agreement in magnitude can be adjusted by a proper choice of the cut-off parameter R_c. If one puts $R_c = \lambda r_c$, where $r_c \equiv (c\hbar/eB)^{1/2}$ is the cyclotron radius, R_c can simply be expressed by a numerical factor λ. In order to fit with the experimental magnitudes of $1/\tau_d$, one has to put $\lambda = 4.0$ for 60° dislocations and 3.7 for 90° dislocations. What causes the difference in λ between 60° and 90° dislocations? One may have to realize here that the 60° dislocation is not a pure edge-type dislocation. We have so far taken b as the magnitude of the Burgers vector for a standard edge-type dislocation, of which only the 90° dislocation is qualified. If we replace, accordingly, b in the 60° dislocation treatment by $b \sin 60° = 0.866b$, then we obtain $\lambda = 3.7$ for the 60° dislocation case. That is in agreement with the 90° dislocation case.

3. *Electron Capture by a Dislocation Line*

We have to make a justification for introducing a cutoff radius R_c in Eq. (11). Physically, this is equivalent to renouncing the existence of electrons within a distance R_c from the dislocation line. Such a requirement will be satisfied by assuming that an electron is captured by the dislocation within this range. Phenomenologically, the capture time of an electron, τ_c, will then be expressed by

$$1/\tau_c = N_d v R_c = N_d v \lambda r_c, \tag{13}$$

where v is the thermal velocity of electron. We have seen experimentally that R_c and hence λ are independent of temperature between 1.5 and 4.2 K. So τ_c is inversely proportional to v; in other words, τ_c, and hence the signal intensity of electron resonance, should be proportional to $T^{-1/2}$. Our observation shown in Fig. 5 strongly indicates the validity of the above assumption. The cyclotron resonance signal in a plastically deformed Ge is quite short-lived. The linewidth is not affected very much by the intensity of illumination. In fact, Eq. (13) yields $\tau_c = 10^{-8}$ s for $N_d = 10^7$ cm^{-2}. This is a value reasonable in understanding the difficulty of obtaining signals. In an undeformed Ge crystal, conversely, we have only $N_d \sim 10^3$ cm^{-2}. That yields $\tau_c \sim 10^{-4}$ s or 100 μs, a typical value of electron lifetime in an ultrapure Ge crystal at liquid helium temperatures.

C. Similarity and Dissimilarity in Cyclotron Resonance Linewidth Treatment between Dislocations and Point Charge Centers

As already pointed out, there is a similarity in the form of scattering potential between point charge centers and dislocations. The former yields an isotropic Coulomb potential while the latter yields a similar form but multiplied by an angular factor. The angular factor in the case of dislocation arises from the anisotropy of the stress field around the edge-type dislocation. Apart from this angular factor and difference in dimensionality, both scattering potentials are of a long-range nature and cause a discrepancy between relaxation times derived by transport theory and by cyclotron resonance linewidth measurement. We shall make a slightly more detailed comparison of these two cases below.

Theory of cyclotron resonance in the long-range scattering potential was first developed by Kawamura et al. (1964) and generalized somewhat later by Miyake (1965). More recently, many calculations appeared in conjunction with the quantum limit treatment, where the condition $\hbar\omega_c \gg k_B T$ prevails (for summary, see Van Royen et al., 1985). The present case, however, can be handled by a more or less classical treatment. So we shall borrow the intuitive idea of the above authors (Kawamura et al., 1964). Consider two extreme cases: i) the motion of an electron is so slow that the phase variation $\Delta\phi$ acquired during the passage through the main force range is much larger than unity, or $\Delta\phi \gg 1$, and conversely, ii) motion of the electron is so fast that $\Delta\phi \ll 1$. In the first case, one arrives at the conclusion that

$$\frac{1}{\tau_d} = \langle \Delta\omega^2 \rangle^{1/2}. \tag{14}$$

This is what we have already assumed to be valid in dealing with the line-broadening in dislocation scattering. In the second case, on the other hand, one obtains

$$\frac{1}{\tau_d} = \frac{1}{\tau_\phi} \langle \Delta\phi \rangle^2, \tag{15}$$

where τ_ϕ is a new parameter supposed to stand for the frequency of phase change.

In applying these ideas to our two-dimensional scattering by dislocations, one has to expect a longer transit time across the force range than in the case of transiting around a point charge center. In addition, variation of the phase angle is expected to be very large near the dislocation line, since the strain field even reverses its sign across the slip plane. These considerations seem to justify our selection of Eq. (14) rather than Eq. (15) for treating the linewidth of dislocation scattering. Even for the three-dimensional scattering case, where less phase change is expected, need for Eq. (15) occurs only above 8 K (Kawamura et al. 1964). Once the situation of Eq. (15) prevailed, we would have the relation $1/\tau_d \propto N_d$ instead of $1/\tau_d \propto N_d^{1/2}$.

Next we shall have an eye on the cutoff radius r_c. Perhaps this parameter is characteristic of the presence of a magnetic field, since the cyclotron motion of electron would become practically impossible within a certain distance from the dislocation line, say λ times the cyclotron radius. A similar cutoff length inherent to cyclotron motion is unavoidable also in the case of ionized impurity scatterings. In an orthodox transport theory, Conwell and Weisskopf (1950) first derived the inverse relaxation time of electron due to collisions with ionized impurities; namely,

$$\frac{1}{\tau_I} = \left[\pi^{3/2} \frac{e^4 N_I}{8m^{*1/2}} k^2 (2k_B T)^{3/2} \right] \ln (1 + \beta T^2) \tag{16}$$

with

$$\beta = \left(\frac{3\kappa k_B}{e^3} N_I^{1/3} \right)^2, \tag{16a}$$

where N_I is the concentration of ionized impurities and κ is the static dielectric constant of the medium. The concentration N_I in Eq. (16a) appears from the cutoff requirement for the force range; that is, to make the maximum range $1/2N_I^{1/3}$. This cutoff is nothing to do with a magnetic field. On application of a magnetic field, electrons start cyclotron motion.

The velocity component of the electron along the magnetic field, which we shall denote v_z, will satisfy the conservation of energy

$$\frac{m^* v_{z'}^2}{2} = \frac{m^* v_z^2}{2} + (n - n')\hbar\omega_c \tag{17}$$

before and after a collision, where n is the Landau quantum number and primes on v_z and n are for the state after the collision. The change in v_z, and accordingly that in wave number k_z, will thus be quantized. The smallest change in k_z is $|\Delta k_z| = \sqrt{2}/r_c$, where $r_c \equiv (c\hbar/Be)^{1/2}$ as before. This value of $|\Delta k_z|$ corresponds to the cutoff radius of $r_c/\sqrt{2}$. If $r_c/\sqrt{2}$ is smaller than $1/2N_I^{1/3}$, N_I in Eq. (16a) should be replaced by $1/2\sqrt{2}r_c^3$. Then the parameter β is no longer dependent of N_I and thus $1/\tau_I$ becomes proportional to N_I. This is what we have observed in electron cyclotron resonance relevent to ionized impurity scatterings in Ge (Otsuka et al., 1973; Otsuka, 1981) for $N_I \geq 5 \times 10^{13}$ cm^{-3}. For $N_I < 5 \times 10^{13}$ cm^{-3}, one finds that the dependence of $1/\tau_I$ on N_I becomes sublinear, almost indicating a dependence of $N_I^{1/2}$. This would be partly due to the reassignment of the cutoff length and partly due to the transition of the phase relation discussed before. All in all, we have seen both similarity and dissimilarity between dislocation and point charge center scatterings associated with the long-range nature of the scattering potential.

III. Natural Potential Wells in ZnSe

A. Cyclotron Resonance in Bulk ZnSe

Despite an intense endeavour to grow a good ZnSe crystal, no attempt to observe cyclotron resonance in this material has ever been found except the one by the present author's group. The reason is not clear; perhaps it is just accidental, since the signal is found to be sharp and beautiful once contemplated (Ohyama et al., 1984a, 1984b).

Small ingots of high purity ZnSe crystal were grown at Tohoku University. The method of growth was a kind of vapor phase transport (Igaki and Satoh, 1979). The first cyclotron resonance sample, cleaved from an as-grown crystal, was nearly oval-shaped and had maximum dimensions of $4 \times 3 \times 0.8$ mm^3. The near bandgap luminescence spectra were taken beforehand and a clear emission peak due to recombination of free excitons was confirmed. It is generally believed that existence of the free exciton luminescence line is an evidence of high purity. Residual impurity concentration was estimated from the photoluminescence feature to be $10^{15} \sim 10^{16}$ cm^{-3}. It turned out later that this estimation was quite reasonable.

1. *Cyclotron Mass of Electron*

Several wavelengths in the far-infrared were taken for observing cyclotron resonance: 172, 220 and 433 μm (1743, 1364 and 692 GHz, respectively) from H_2O-, D_2O-, and CO_2-pumped lasers, respectively. These were all operated in pulses at a frequency of 30 Hz in synchronized combination with the photoexcitation pulses by a xenon flash lamp. Since the $\omega_c\tau$ value turned out to be so high, millimeter wave measurements at 35 GHz later took the initiative for detailed study of the properties of the electron system in ZnSe. For the FIR measurements, experiments were carried out in the Faraday configuration and the magnetic field generated by a superconducting solenoid was always vertical to the (110) plane on which both the excitation light and the far-infrared beam were incident.

In Fig. 6, a typical 1364 GHz trace for an as-grown ZnSe crystal is shown. Two distinct peaks near 7 T are obvious and reproducible. One is associated with the effective mass of $0.145m_0$ while the other $0.149m_0$,

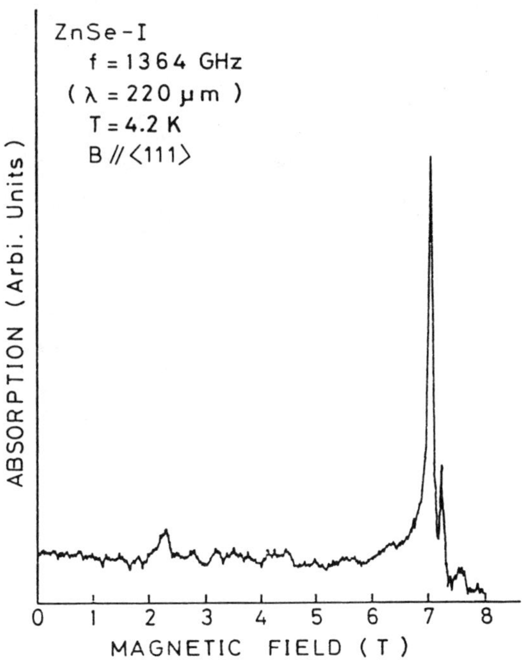

FIG. 6 Far-infrared cyclotron resonance trace from a high-purity as-grown ZnSe crystal. Two sharp peaks are observed, corresponding to cyclotron masses of $0.145m_0$ and $0.149m_0$. The signals are observable only right after the intrinsic photoexcitation pulse (Ohyama et al., 1984a).

where m_0 is the free electron mass. It seems that the resonant peaks are based on another Drude type absorption, though not very clearly evidenced. The resonant absorption peaks are quite short-lived. At 4.2 K, they completely disappear in a few microseconds after the end of photopulse.

With the help of 35 GHz waves in combination with an electromagnet producing a rotatable horizontal magnetic field, the $m^* = 0.145m_0$ peak is found to be isotropic. This is shown in Fig. 7. The measurement is at 1.5

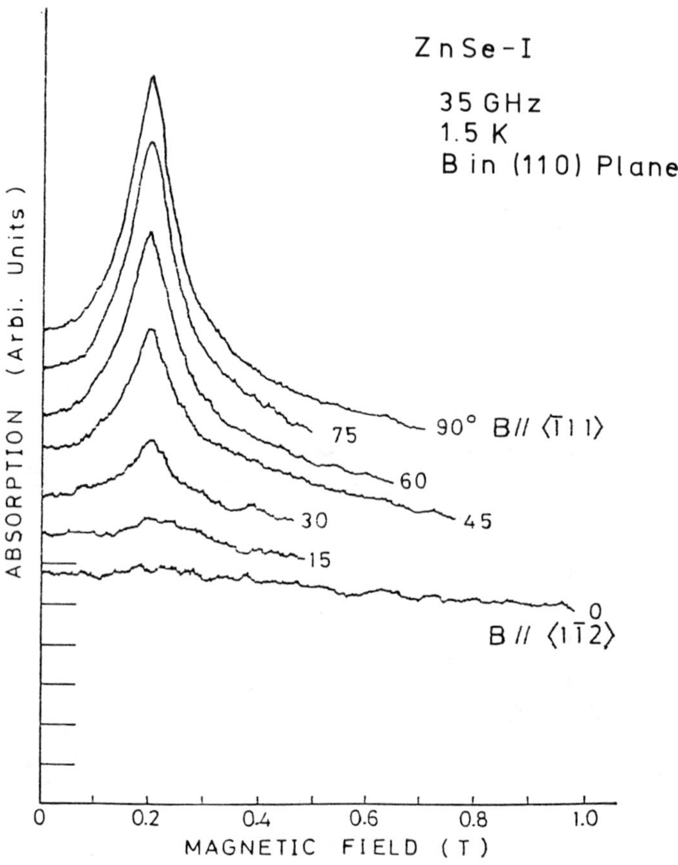

FIG. 7 Millimeter wave cyclotron resonance traces from the same sample as used for Fig. 6. Each trace was taken 2 μs after the photopulse. The angles indicated are between the magnetic field and a characteristic axis ⟨112⟩. Anisotropy of the signal height is to be noted, the signal completely vanishing at B ∥ ⟨112⟩ and becoming maximum at B ∥ ⟨111⟩. The peak is always on a nonresonant background. The baseline for each trace is shown along the ordinate axis in the ascending order of the angle from the characteristic axis (Ohyama et al., 1984a).

K. A peculiar feature should be discussed. The resonant peak is isotropic but its intensity strongly depends on the orientation of the magnetic field that is rotated within the (110) plane. The angle 0° corresponds to the magnetic field direction along $\langle 1\bar{1}2\rangle$ and 90° to $\langle \bar{1}11\rangle$. Evidently the anisotropy has a two-dimensional character. Detailed features of this two-dimensionality will be discussed in B.

Though there is no cyclotron resonance observation available elsewhere, perhaps the most reliable determination of the effective mass was that of Hölscher, Nöthe and Uihlein (1985) by means of an interband two-photon magnetoabsorption. They yield $m_e^* = 0.147m_0$ within an accuracy of 2%. That is in a quite good agreement with our result of $m^* = (0.145 \pm 0.002)m_0$, and it offers us a substantial support that we are observing an electron signal.

As for the smaller peak at $m^* = 0.149m_0$, we have at present no conclusive identification. In 35 GHz measurement, it is not resolved. One possible speculation would be a polaron effect. In fact, a theoretical calculation as suggested by Bajaj (1970) with $\alpha = 0.432$, which is the Fröhlich coupling constant calculated by Rode (1970), predicts a second peak at $m^* = 0.148m_0$ and a third at $m^* = 0.153m_0$. This prediction is in agreement with our observation of the second peak so far as the resonance position is concerned. However, the observed relative intensities of the first and second peaks require an optically hot state of the photoexcited conduction electrons in order to justify the polaron model. Be that as it may, the ordinary Fröhlich relation $m^* = m_e^*(1 + \alpha/6)$ yields a band mass value of electron m_e^* to be $0.135m_0$. The so-called band mass reported by Hölscher et al., accordingly, should be accounted for by the polaron mass.

2. *Transport Argument*

As already mentioned in II, we can convert the observed linewidth of cyclotron resonance absorption into the inverse relaxation time $1/\tau$ and hence the electron mobility $\mu_e = e\tau/m^*$. We find $1/\tau = 5.3 \times 10^{10}\ \text{s}^{-1}$ at 4.2 K and this depends little on temperature down to 1.5 K. It thus seems that at liquid helium temperatures the electron relaxation time is mainly governed by neutral impurity scatterings. The cyclotron mobility at $B = 7\text{T}$ is found to be $2.3 \times 10^5\ \text{cm}^2/\text{Vs}$. This is indeed an incredibly high value for II–VI compound semiconductors. Tentatively neglecting the contribution from lattice scatterings and assuming that the so-called Erginsoy relation (Erginsoy, 1950)

$$\frac{1}{\tau_N} = \frac{20N_D\hbar a_B^*}{m^*} \tag{18}$$

is valid, where τ_N is the electron collision time with neutral donors, N_D the concentration of the neutral donors and a_B^* the effective Bohr radius of the donor impurity, we find $N_D \sim 10^{15}$ cm^{-3}. Here we have assumed $a_B^* = 33$ Å in accordance with an electron mass of $0.145m_0$ and the Bohr radius expression of $a_B^* = \kappa\hbar^2/m^*e^2$, with κ being the static dielectric constant. This is indeed a very crude argument, since we are simply assuming that impurities are all donors. Of course there must also be acceptors. But their contribution to the cyclotron resonance linewidth will be small in comparison with that of donors. The relative unimportance of acceptor scatterings has been evidenced already in Ge and Si (Otsuka et al., 1964, 1966, 1968; Otsuka, 1984). The essential reason is that the electron-neutral donor scattering is to be compared with the electron–hydrogen atom (e^--H) scattering while the electron-neutral acceptor scattering with the positron–hydrogen atom (e^+-H) scattering. The cross section of the e^--H scattering is much larger than that of the e^+-H scattering at all wave numbers of the incident particle. So we shall forget all about acceptors so long as impurity scatterings are concerned.

Temperature was changed up to 77 K and the linewidth was measured as a function of temperature for 35 GHz. Derivation of the relaxation time was possible almost up to ~40 K. At higher temperature, broadening of the line made the resonance more or less of the Drude type. The broadening is considered to arise from acousto-piezoelectric and optical phonon scatterings. The inverse relaxation times due to acousto-piezoelectric and optical phonons are given by

$$1/\tau_{pz} = 7 \times 10^{13} \left(\frac{m^*}{m_0}\right)^{1/2} T^{1/2} \left(\frac{K^2}{\kappa}\right) \text{ s}^{-1} \tag{19}$$

and

$$1/\tau_{op} = 2\alpha\omega_0 \exp\left(\frac{-\Theta}{T}\right), \tag{20}$$

respectively, where K is the piezoelectric electromechanical coupling constant, κ the static dielectric constant, ω_0 the optical phonon frequency and Θ the Debye temperature. Contribution from the acousto-deformation potential scattering is much less important. In addition to the contributions in Eqs. (19) and (20), one has to consider the contribution from impurity scatterings. In Fig. 8, we give the temperature dependence of the total observed inverse relaxation time. The dashed lines numbered 1, 2 and 3 give the contributions from optical phonon, acousto-piezoelectric phonon and impurity scatterings, respectively. The phonon scattering contributions are independent of the sample, while the impurity scattering contribution is expected to vary from sample to sample. In the present case, the impurity scattering is calculated in terms of the Erginsoy formula in Eq. (17), putting $N_D = 1 \times 10^{15}\ \mathrm{cm}^{-3}$. Other parameters employed are $\omega_0 = 4.6 \times 10^{13}\ \mathrm{s}^{-1}$, $\Theta = 354$ K, $m^*/m_0 = 0.145$, $K = 0.0437$, $\kappa = 8.7$ and $a_B^* = 33$ Å. The solid line in Fig. 8 is a result of summation over the three contributions mentioned. It is qualitatively in agreement with experimental points, though some discrepancy is obvious at higher temperatures. The reason for the discrepancy is not clear. Perhaps we have put too much confidence on the existing values of the physical parameters. It seems, however, that very little contribution is coming out of ionized impurity centers, though the contribution from ionized impurity centers to the cyclotron resonance linewidth does not necessarily mean a broader linewidth at lower temperatures (Otsuka et al., 1973).

B. Two-Dimensional Behavior in 35 GHz Observation

As already indicated in Subsection IIIA, there exists a spectacular two-dimensional characteristic in cyclotron resonance. This is prominent in 35

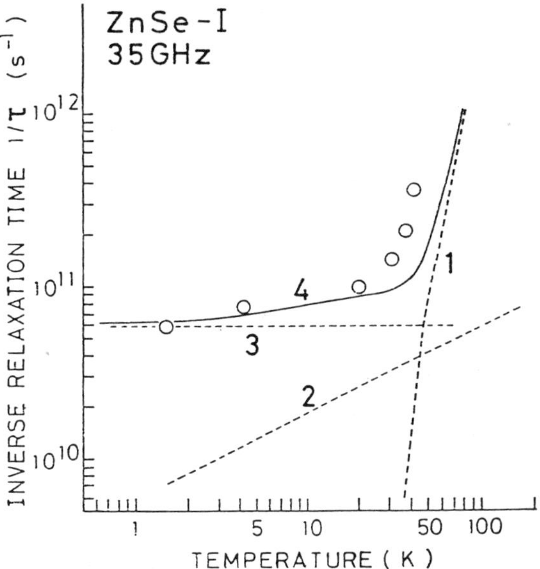

Fig. 8 Temperature dependence of observed and calculated inverse relaxation times of electrons in a high-purity ZnSe crystal. Open circles are experimental points obtained by 35 GHz cyclotron resonance. Dashed lines 1, 2, and 3 are for calculated polar optical phonon scatterings, acoustopiezoelectric phonon scatterings, and neutral donor scatterings with $N_D = 1 \times 10^{15}\ \mathrm{cm}^{-3}$, respectively. The solid line 4 gives the total contribution from 1 through 3 (Ohyama et al., 1984b).

GHz observation. Figure 7 shows a strong anisotropy in absorption intensity. On top of this anisotropy, one may note another peculiarity in Fig. 7; existence of a nonresonant background. Along the ordinate, a series of horizontal bars are shown. They are to indicate the zero of each trace. The zero of the 0° trace coincides with the abscissa and the correspondence is in the ascending order with angle for the rest of the traces. So we have two things unexplained. One is the reason for the anisotropy of the intensity of the resonant peak and the other is the origin of the background.

Time resolution measurement for the background shows that this is quite long-lived at low temperatures in comparison with the resonant signal peak. On raising temperature, however, the lifetime of the nonresonant signal drops. The nonresonant signal can naturally be observed at zero magnetic field in the form of microwave photoconductance. The decay profile of the microwave photoconductance signal with temperature as the varying parameter is obtained at zero magnetic field and shown in Fig. 9. This profile gives credit to a supposition that carriers responsible

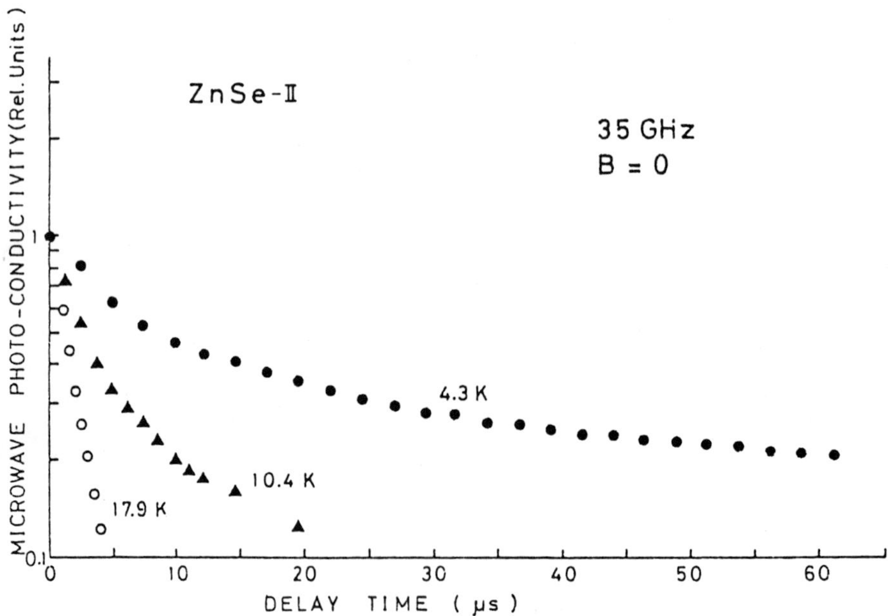

FIG. 9 The decay profile in time of the nonresonant background absorption, obtained as the 35 GHz photoconductance signal at zero magnetic field. The time constant varies with temperature, being larger at lower temperature. General features in the decay process are more or less the same for all samples.

for the background signal are localized ones that require a certain thermal energy to get free and recombine with oppositely charged carriers.

Both the anisotropy in signal strength and the appearance of the background depend on the sample. By taking another as-grown sample, one finds a different character in the background signal. This is shown in Fig. 10. The background signal is no longer flat, going up steeply as the magnetic field is reduced. Moreover, the slope is somewhat anisotropic. The resonant signal is small in comparison with the background but less anisotroplic in intensity than in the other crystal. To see the reason for zero field elevation in the background signal, we have raised the ambient temperature to 20 K. A striking feature shows up as seen in Fig. 11. The background can also have a resonant peak! And this peak is highly anisotropic. In Fig. 12, the angular dependence of the resonance magnetic field, equivalently an apparent effective mass, is shown both for the original

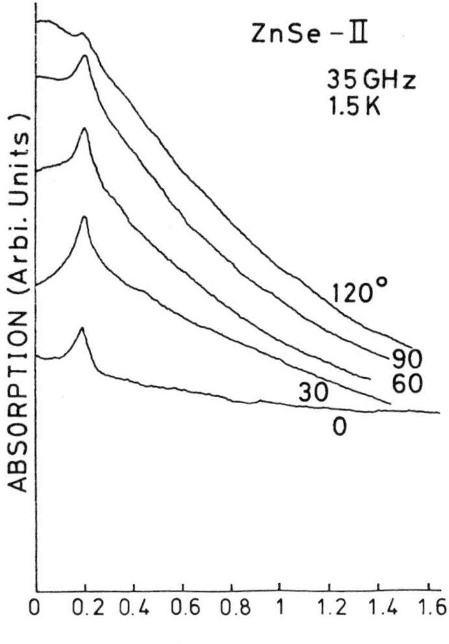

FIG. 10 Appearance of the nonresonant background absorption seems to change from sample to sample. In this ZnSe-II sample, elevation at low magnetic fields is observed for certain orientations of the external magnetic field. One may also note that anisotropy in intensity of the resonant absorption peak is not so large as that for ZnSe-I.

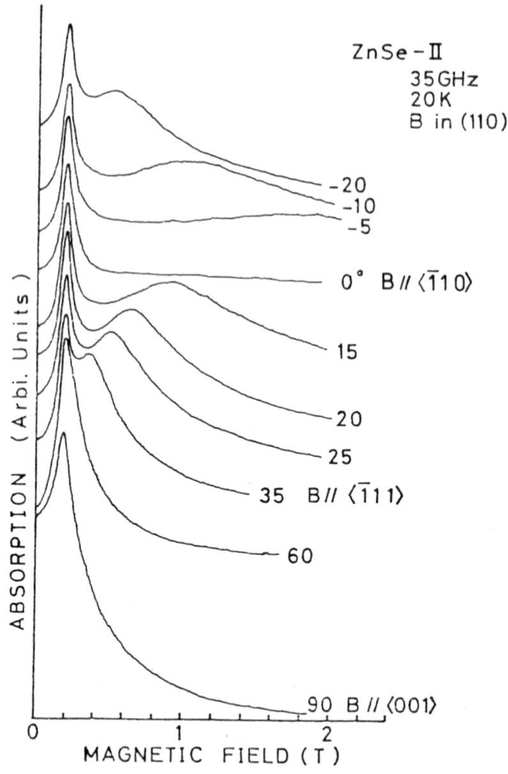

FIG. 11 At an elevated temperature of 20 K, the background absorption of ZnSe-II becomes resonant. Beside the isotropic resonance peak appearing at $\sim 1\ T$, one can see the second peak that greatly changes its position in the magnetic field with the direction of the external magnetic field.

resonant peak and for the background. The former is isotropic as before, while the new peak arising from the background has a distinct two-dimensional anisotropy. The apparent effective mass for the latter diverges at $B \parallel \langle\overline{11}2\rangle$ and coincides with that of the isotropic resonance at $B \parallel \langle 111\rangle$. The coincidence of two effective masses at $B \parallel \langle 111\rangle$ shows an evidence that the two-dimensional system is arising from electrons, if the $m^* = 0.145m_0$ resonance is electronic. Our task now is to give an explanation for the two-dimensionality and for the coexistence of two- and three-dimensional electrons (Ohyama et al., 1985).

A microscopic observation with optical polarizers shows distinct stripes. They are spaced at a distance of $100 \sim 200\ \mu$m. Thanks to a subsidiary X-ray analysis, it has turned out that the stripes are twin crys-

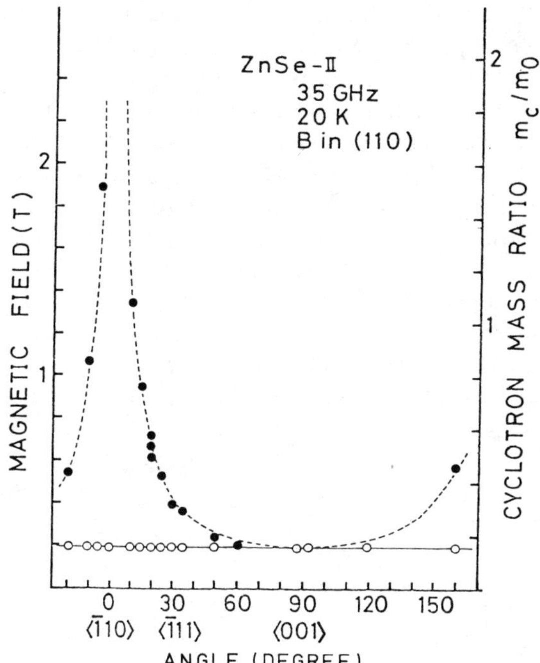

FIG. 12 The peak position in magnetic field, or the apparent cyclotron mass of the background signal obtained from ZnSe-II at 20 K, is plotted against the direction of the magnetic field. The resonance field coincides with that of the isotropic resonance for B ∥ ⟨001⟩, and evidently the resonance shows a two-dimensional nature.

tal boundaries, vertical to the direction of the crystal growth, ⟨111⟩; in other words, they are aligned along ⟨112⟩. These stripes almost always appear in the present method of crystal growth. After this visual observation of twin crystal boundaries, we are naturally inclined to associate the observed two-dimensionality of the resonance with the localization of electrons along the interfaces. A more definite search for the evidence for a two-dimensional localization will be given below.

C. Electron Accumulation Layers at Crystal Boundaries

1. *Dependence on the Sample Treatment*

It is of interest to see a variation in appearance of the background signal depending on the treatment of an as-grown crystal. The idea is to see the effect of changing the electron density. In addition to the observation made on as-grown crystals, we have tested two other samples: one is the so-called zinc-dip crystal, that is a crystal immersed in the melt of zinc metal for 24 hours, and the other is a crystal treated in a saturated vapor atmosphere of selenium. Evidently the former process acts to fill up the zinc vacancies, releasing more electrons into the conduction band, while the latter process increases zinc vacancies that accept more electrons.

The effect of "zinc-dip" is striking. "Free carriers" are now available even without photoexcitation at liquid helium temperatures. But these "free carriers" are two-dimensional. On raising temperature, say to 18 K, one can also see the three-dimensional resonance on top of the broader two-dimensional resonance. The feature is shown in Fig. 13. It thus seems that the two-dimensional electron system is almost degenerate. The free carrier density, which is responsible for the cyclotron resonance peak, is nearly unchanged with temperature for the two-dimensional system. Only

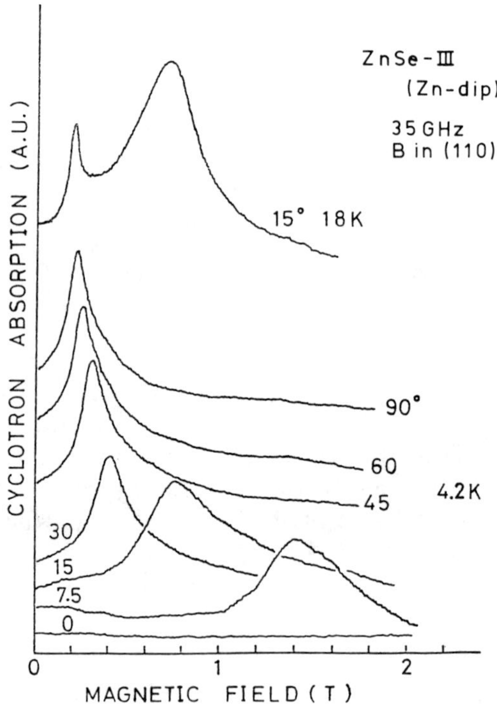

FIG. 13 Resonance feature in a "zinc-dip" crystal. Two-dimensional anisotropic resonance shows up even at 4.2 K and even without photoexcitation. In other words, the two-dimensional electron system seems to be nearly degenerate. No isotropic resonance is seen at 4.2 K. On raising temperature, the three-dimensional isotropic resonance signal also appears as typically shown at the top, for 15° and at 18 K.

the electron density of the three-dimensional system changes drastically with temperature.

The effect of selenium vapor treatment is rather depressing. Nothing but a flat nonresonant signal appears at 4.2 K even with photoexcitation. Only at an elevated temperature, say at 26 K, does an isotropic resonance signal appear. Evidently this is due to the three-dimensional electron system. The resonant signal is quite short-lived and appears only right after the photoexcitation pulse.

What is causing these effects? The answer has practically been given in our preceding statements. Electrons can be accommodated along the twin crystal boundaries. They form a two-dimensional system. Such a mechanism has already been pointed out by Landwehr and Uchida (1985) for the crystal boundaries of p-type Ge. In their case, the accommodated electrons are those of an inversion layer. On the other hand, our system in ZnSe is building up an accumulation layer, since the bulk material is always n-type.

The effect of "zinc-dip" is to raise the overall electron density. The number of electrons accommodated in each crystal boundary is also raised. Their Fermi level exceeds the mobility edge and electrons can move freely in the boundary region even at 4.2 K. This is the reason why the two-dimensional peak appears at 4.2 K even without photoexcitation. On elevating temperature, a certain fraction of electrons in the accumulation layers will spill out of the boundaries and start cyclotron motion outside, thus giving rise to the three-dimensional resonance.

The effect of selenium vapor treatment, on the contrary, is to reduce the electron density. Unless one illuminates the crystal with bandgap light, scanty electrons are deeply hidden in the localized states along the crystal boundary sides. Even photoexcitation gives rise to only a nonresonant signal at 4.2 K. The photoelectrons are promptly absorbed by the crystal boundaries, where electrons cannot exceed the mobility edge. Raising the temperature finally helps to produce some unlocalized electrons.

The case of a good as-grown crystal is intermediate between zinc-dip and selenium vapor treated cases. We usually have both resonant and nonresonant signals at liquid helium temperatures. The resonant one is from three-dimensional electrons and the nonresonant one from two-dimensional electrons. Neither one will show up without photoexcitation. At elevated temperatures, the nonresonant signal can become resonant, giving rise to an anisotropic appearance of the resonance peak. This is coexistent with an isotropic resonance peak coming from the three-dimensional electron system. The relative signal strength between two- and three-dimensional electron systems differs from sample to sample, depending partly on the density of the crystal boundaries and partly on the electron mobility within the bulk region. Dependence on the electron mobility may sound strange at first. But that is true. Free carriers are first produced in the bulk region. Then, electrons will be drawn to the interfaces where they can find more stable accommodation. If the electron mobility is low in the bulk, or perhaps if electrons repeat capture and reemission processes there, the electron system as a whole will remain for a long time in the three-dimensional region. On the contrary, if the mobility is high, electrons will rush to the interfaces without being obstructed. As a matter of fact, by slicing off a part filled with crystal boundaries, we have obtained an enormous enhancement of the three-dimensional electron signal in an extremely high-mobility sample. Substantially losing two-dimensional accommodation, electrons have but to remain three-dimensional with a long life.

Finally, mention should be made of the anisotropy of the three-dimensional resonance intensity. A most reasonable explanation for this phenomenon will be to admit the existence of layers of small-scale stacking faults. A large-scale stacking fault may grow into a twin crystal interface which we can observe visually. Invisible stacking faults do not change the crystalline orientation but will capture electrons. If these stacking faults are aligned parallel, and vertical to the direction of crystal growth, one can understand our experimental observation. The average separation between two stacking faults would be of the order of the cyclotron radius at 35 GHz, or perhaps larger than the cyclotron radius by a factor comparable to λ that was introduced in Section II in conjunction with the cutoff radius around a dislocation line. With a separation of at most a few microns, these stacking faults are aligned between crystalline interfaces, which are separated by 100 $\sim$ 200 microns. These features are schematically illustrated in Fig. 14.

IV. Deep-Level Defects in Semi-Insulating GaAs

Defects in semi-insulating GaAs are drawing special attention because of their possible roles, either good or evil, in future integrated circuits based on this material. Of all the imperfections, particularly controversial is the one called EL2. The nature of this defect has been widely traced and discussed much since its first discovery (Lin et al., 1976; Bois and Vincent, 1977; Martin et al., 1977 etc.). Experimental approaches have mostly been either from photoluminescence or from DLTS (deep-level transient spectroscopy). Only recently, microwave and far-infrared absorption measurements have been introduced. The former has been playing a role in identification of EL2 (Elliot et al., 1984; Baumler et al., 1985;

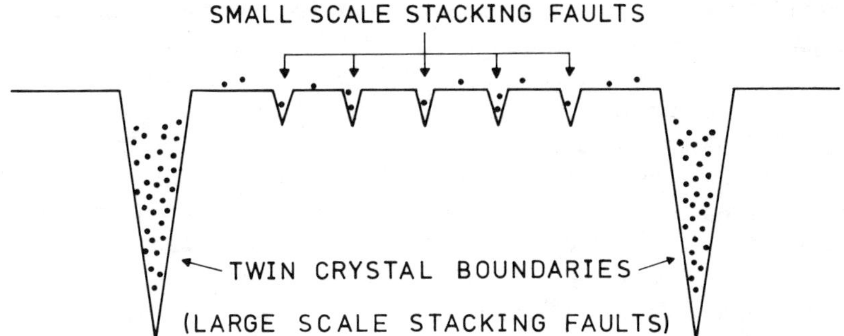

FIG. 14 Schematic view of stacking fault potentials responsible for the anisotropy of cyclotron resonance. The small-scale stacking faults are considered to cause the anisotropy in intensity of the three-dimensional electron resonance. The large-scale stacking faults, on the other hand, grow into their crystal boundaries that can accommodate a two-dimensional electron system. Dots indicate electrons.

von Bardeleben et al., 1985), while the latter has provided a model for the metastable state of EL2 (Ohyama et al., 1985, Otsuka et al., 1985). The present author is involved in the latter, so he will refer to the identity of EL2 within the range of consistency with his far-infrared results.

It has been suggested that EL2 is something connected with the so-called antisite As, in other words, As atom (or atoms) on the site (or sites) of Ga (Elliot et al., 1984; Baümler et al., 1985). It behaves like a deep donor, photosensitive but loses its photoresponse on mid-gap photoexcitation at low temperatures. This loss of photoresponse is called photoquenching and is regarded as the most direct fingerprint of EL2. The photosensitivity is recovered by heating the sample, say above 100 K.

A. Photoquenching in Donor Zeeman Transition

1. *General Characteristics*

Access to EL2 in semi-insulating GaAs crystals by the far-infrared measurement has been induced in quite an unexpected way. We take two undoped samples, one as-grown and the other ingot-annealed at 800°C for 24 hours, and another In-doped sample. All of them are LEC (liquid encapsulatd Czochralski) grown crystals. They have a size of $4 \times 4 \times 0.5$ mm^3. What we measure is the absorption intensity of the donor Zeeman transition. That is $1s$ to $2p_{+1}$ transition of an electron bound to a shallow donor impurity. Even in an undoped sample, there always exist a measurable amount of neutral donors. The donor Zeeman absorption shows up, especially when the sample is illuminated by the band-gap light. The bandgap light produces a number of free carriers, electrons and holes. They promptly neutralize the ionized impurities. Far-infrared absorption by neutral donors (and acceptors) can thus be observed. This is a usual feature in Zeeman transitions in III-V compound semiconductors (Ohyama and Otsuka, 1983). The total concentration of shallow impurities can be estimated by the FIR Zeeman transition measurement with the help of photoexcitation (Ohyama, 1982; Ohyama et al., 1982; Ohyama and Otsuka, 1983). Using the absorption cross section obtained for the $1s$ to $2p_{+1}$ Zeeman transition (Ohyama, 1982), we have derived the concentration of genuine shallow donors, say Si or S, typically in the ingot-annealed LEC-grown material, to be approximately 5×10^{13} cm^{-3}. A new feature, however, is to be introduced here. We shine light on the crystal with a xenon flash lamp that emits spectra ranging from mid-gap to above-gap energies. By inserting an infrared filter in the optical path, one can reduce the excitation by the above-gap light. The mid-gap excitation is unaffected. Then one can observe a "photoquenching" of the donor Zeeman transition by the mid-gap excitation.

In Fig. 15, we plot the intensity of the donor Zeeman absorption for the above three samples as a function of time after starting the illumination by the IR-filtered xenon flash lamp that is operated at 15 Hz with a pulse width of 1 μs. The far-infrared source is an H_2O vapor laser operated in pulses at 30 Hz and at a wavelength of 119 μm (2521 GHz). As is always the case in our photoexcitation measurement, the FIR laser pulses are in synchronized combination with the photoexcitation pulses. The measurement is carried out at 4.2 K. It is seen in Fig. 15 that the Zeeman absorption signal intensity decreases with the passage of time, the time rate of decrease is different for each sample. The signal is most long-lived for the undoped ingot-annealed sample and most short-lived for the In-doped sample. For all the three samples, at any rate, the signal is quenched after a certain lapse of time. As already stated, this type of quenching, associated with mid-gap excitation, is the fingerprint of the so-called EL2, though we have not expected that to occur for the far-infrared donor Zeeman absorption. Evidently, EL2 is acting here, but how? What we observe is the Zeeman transition of a shallow donor. But we are sure that the shallow donor itself is not EL2, since we have never observed a photoquenching of the shallow donor Zeeman transition signal elsewhere. The photoquenching corresponds to the formation of a metastable state of EL2. So long as EL2 remains photosensitive, our shallow donor impurities do not lose their Zeeman transition signals. It is when EL2 is made optically inactive that the Zeeman absorption is quenched. All we can be sure so far about EL2 is that it is a deep donor in its optically active state.

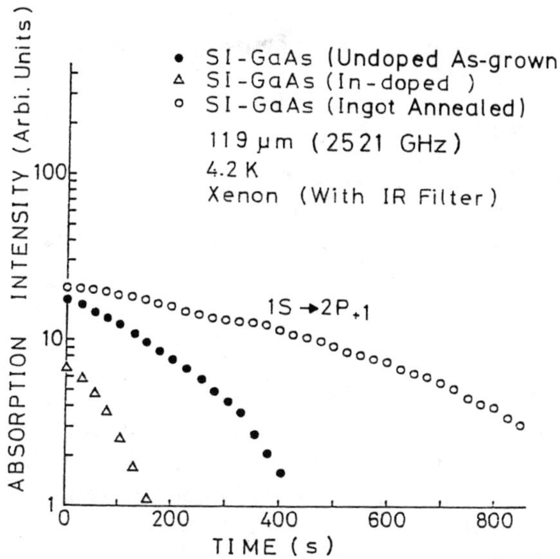

FIG. 15 Photoquenching time profiles of the far-infrared Zeeman absorption by shallow donors for three samples as indicated, one doped and two undoped with and without annealing. Use of the infrared filter enhances the role of the mid-gap (extrinsic) excitation, prompting the occurrence of photoquenching. The abscissa gives the time lapse after starting the 15 Hz photopulse excitation (Ohyama, Nakata, and Otsuka, 1985).

Be that assumed to be neutral. After photoquenching, then, it becomes singly ionized. This ionized defect may, after a possible voyage for a distance, accept an electron from a neutral shallow donor and share it with the ionized donor. Can that not be a metastable state of EL2, in association with a shallow donor impurity? Indeed this is a model for the metastable state of EL2 proposed by the author's group (Otsuka et al., 1985). In the usual thermal equilibrium state of a semi-insulating GaAs, almost all the shallow impurities are in ionized states even at liquid helium temperatures. Indeed, that accounts for the semi-insulation; otherwise provision of electrons from the neutral shallow donors into the conduction band would make the material semiconducting rather than semi-insulating. In such a material, a deep donor called EL2, however, is neutral. The usual photoquenching observed after the mid-gap photoexcitation would then be re-sharing of the electron that was originally bound to EL2 with an ionized donor center. If both shallow donors and deep donors (EL2) remain neutral, one can never observe a photoquenching. This can be confirmed by illuminating the semi-insulating GaAs crystal with above-gap light. In fact, monochromatic excitation by an argon ion laser at the wavelength of 514.5 nm does not bring forth any photoquenching. Such an illumination simply helps neutralization of impurities.

2. *Formation of the Metastable State*

Putting the identity of EL2 aside for the moment, let us denote the neutral state of the defect by $X^{\cdot}$, with the superfix dot standing for the bound electron. The singly ionized state will then be expressed as X^{+}. The process of the transition to the metastable state will be expressed by a reaction formula

$$X^{\cdot} + D^{+} \rightarrow (X^{+}D^{+})^{\cdot} \tag{21}$$

where D^{+} denotes the ionized shallow donor state. Frequently the metastable state of EL2 has been expressed as EL2*. This is optically inactive and considered to be displaced in the configuration space. What we do here is to identify it as molecular hydrogen in H_2^+-like structure; two positive ions separated by a certain distance and sharing an electron. Under photoexcitation by a xenon lamp, the situation will be a little bit more complicated. The light contains both intrinsic and extrinsic parts. The intrinsic component neutralizes both EL2 and shallow donors while the extrinsic component ionizes them. Then we shall have $X^{\cdot}$, X^{+}. $D^{\cdot}$ and D^{+}. Only the combination of Eq. (21) and/or

$$X^{+} + D^{\cdot} \rightarrow (X^{+}D^{+})^{\cdot} \tag{22}$$

will lead to the metastable state. Insertion of the infrared filter, which cuts off a large part of above-gap energy photons, favors the combination in Eq. (21) more, since more of the shallow impurities remain ionized. To be more exact, association of shallow and deep donors is expected to occur via an electron in the continuum; in other words, perhaps we had better write

$$X^{+} + e^{-} + D^{+} \rightarrow (X^{+}D^{+})^{\cdot} \tag{23}$$

instead of Eq. (21) or Eq. (22). Here e^{-} is an electron in the conduction band. It goes without saying that the electrons, free in the conduction band, yield a cyclotron resonance signal. A typical feature is demonstrated in Fig. 16. Here we have two signals for the as-grown semi-insulating GaAs: a very broad absorption line arising from the donor Zeeman transition and another broad line arising from the cyclotron transition of conduction electrons. These signals have been obtained without using an IR filter and were taken in time resolution right after the xenon flash lamp excitation pulses. The far-infrared wavelength is 220 μm (1364 GHz). For comparison, signals obtained from an n-type MBE (molecular beam epi-

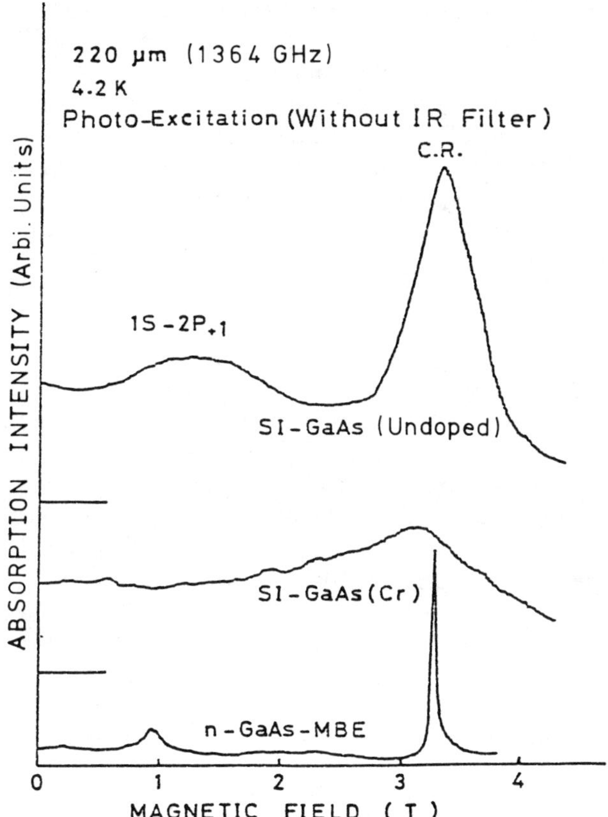

FIG. 16 Far-infrared cyclotron resonance and Zeeman absorption signals, taken at 4.2 K, right after the unfiltered xenon flash lamp excitation pulse. Two GaAs samples, one doped and the other undoped, are semi-insulating and yield broad resonance absorption signals. For comparison, absorption signals obtained from an MBE-grown sample are shown at the bottom (Ohyama, Nakata, and Otsuka, 1985).

taxy) grown GaAs are also shown. One may note how much the signals in the semi-insulating crystal are broadened in comparison with those in the MBE crystal. The broadening of the cyclotron resonance signal is no doubt arising from the enhanced collision frequency of electrons with impurities.

3. *Effect of Annealing*

We shall next pay attention to the difference between as-grown and ingot-annealed samples. These two crystals, though undoped, originally contained nearly the same amount of shallow donor impurities. In Fig. 17, we show two traces of far-infrared magnetoabsorption: one for as-grown and the other for ingot-annealed samples. Both Zeeman and cyclotron resonance signals are shown. These signals are obtained more than 1,000 seconds after the start of the experiment, or after starting illumination by a xenon lamp without an IR filter. At the very beginning, signals including those of cyclotron resonance are enormous, but they are rapidly quenched while sweeping the magnetic field, so that no stationary signals can be obtainable. Only after waiting approximately 1,000 s can relatively stabilized traces as shown in Fig. 17 be obtained. In other words, signals in Fig. 17 are seen after a considerable progress in photoquenching. Yet one can see a considerable difference between two crystals. The effect of photoquenching is larger for the as-grown crystal than for the ingot-annealed one. This fact is consistent with the generally accepted idea that annealing is effective in reducing the number of EL2 centers.

One may ask why the traces in Fig. 17 have been obtained without using an IR-filter. By using an IR-filter, the effect of photoquenching would certainly be enhanced. That is true, but the effect is too much. After the same lapse of time, i.e. 1,000 s, all of the signals die down for both crystals, so that no comparison can be made. The unfiltered light of a xenon lamp is just suitable for the present purpose to see the difference between as-grown and annealed crystals.

B. Connection with Photoluminescence Observation

The EL2 complex has drawn attention on account of its peculiar behavior in photoluminescence. Accordingly, it becomes necessary to examine correspondence between our far-infrared observations and photoluminescence properties for the same samples. The so-called photoquenching effect has so far been observed for the mild-gap transition (Tajima, 1984). In our measurement, we observe another quenching behavior in a near-gap transition.

In Fig. 18, we present steady-state photoluminescence traces for two undoped crystals, one as-grown and the other ingot-annealed, obtained at 4.2 K. The samples are exposed to continuous illumination by the intrinsic (2.41 eV) light of an Ar^+ laser. We find a sharp photoluminescence peak at a photon energy of 1.492 eV. That is very close to the bandgap energy, and is supposed to be the conduction band-to-acceptor, or rather the donor-to-acceptor, level transition. On the left of each trace, a difference in the size of the signal is indicated. For the lower trace, the signal size has been reduced by a factor of 10. As a matter of fact, a factor of 10 difference exists between the upper signal for the as-grown crystal and the

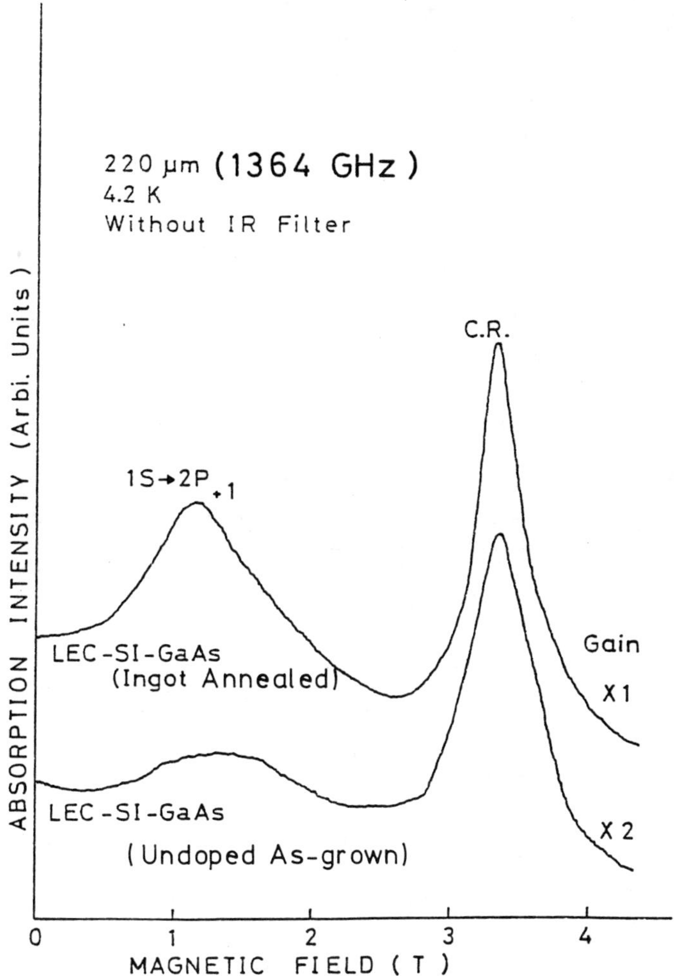

FIG. 17 Far-infrared cyclotron resonance and Zeeman absorption signals, taken right after the excitation pulse in time-resolution but nearly 1,000 s after starting the experiment, are shown for two undoped semi-insulating GaAs crystals, one as-grown and the other ingot-annealed. These signals are obtained without using an IR-filter. If the filter is inserted, the entire signals disappear for both samples after the same time-lapse of 1,000 s. Only an unfiltered photoexcitation can yield such a comparison between the two samples. Evidently, the annealing process is reducing the photoquenching effect (Ohyama, Nakata, and Otsuka, 1985).

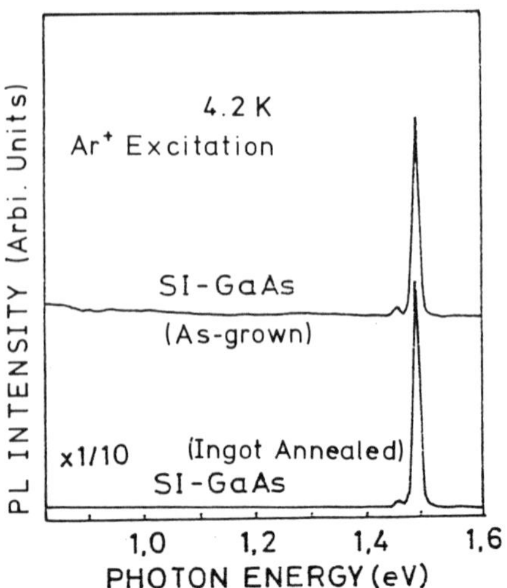

FIG. 18 A near-gap (1.492 eV) photoluminescence spectrum under steady illumination by an intrinsic light from an argon ion laser is compared to 4.2 K between two undoped semi-insulating GaAs crystals, as before, one as-grown and the other ingot-annealed. The emission is stronger for the annealed sample by an order of magnitude.

lower signal for the ingot-annealed crystal. As already stated, these two crystals originally had almost equal amounts of impurities. So such a big difference as a factor of 10 in the photoluminescence intensity does not arise from the difference in impurity concentration, but from the action of heat treatment.

What we want to stress here, however, is not the effect of heat treatment on the photoluminescence intensity, but something connected with the behavior of photoquenching. So long as one continues the steady photoexcitation by an Ar^+ laser, the observed photoluminescence signal at 1.492 eV does not change its intensity both for as-grown and for ingot-annealed crystals. But once we 1) quit the illumination by the Ar^+ laser, 2) expose the sample to an unfiltered xenon flash lamp for three minutes and 3) cut the xenon lamp light and resume illumination by the Ar^+ laser, some peculiarity arises. The feature is shown in Fig. 19, where the 1.492 eV photoluminescence intensity is plotted against the time passage after resuming the Ar^+ laser illumination. For the as-grown sample, the photoluminescence is quenched at the beginning, but gradually recovers its original strength with time. For the ingot-annealed sample, on the other hand, the photoluminescence is first enhanced instead of being quenched, and then gradually decays to the steady-state intensity. The apparent steep rise for the ingot-annealed sample is perhaps not real, since we

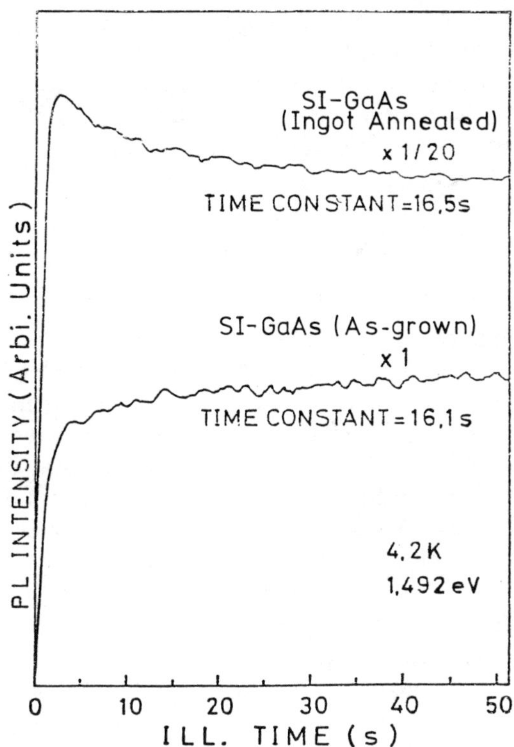

FIG. 19 Time variation of the near-gap (1.492 eV) photoluminescence intensity for two undoped semi-insulating GaAs crystals, as-grown and annealed. The crystals have been illuminated for three minutes by unfiltered xenon flash lamp to cause the photoquenching of the 1.492 eV line. Then one starts an intrinsic illumination by an argon-ion laser to recover the original intensity of the same line. The abscissa gives the illumination time by the argon ion laser after shutting off the xenon light. Recovery time constants are nearly the same, but an initial signal enhancement at the annealed sample is to be noted.

obtain the signal after a slow response lock-in amplifier. So what we can be sure of concerning the annealed sample, is the photoenhancement, in place of the photoquenching for the as-grown sample. The reason for this photoenhancement is quite unexplained at present, although a certain amount of speculation about photoquenching is available. We have seen, at any rate, that some photoquenching as well as photoenhancement effects are observed in photoluminescence spectra for the same samples in which the far-infrared quenching in the donor Zeeman signal is observed. The recovery of the effect in the illumination by the Ar^+ laser light is due to the intrinsic nature of the illumination. The prior illumination by the

Ar^+ laser, before exposure to the xenon lamp light, does not cause any photoquenching. In the course of three minutes of exposure to the xenon lamp, the samples experience the extrinsic part of the illumination, and certainly the cause of photoquenching is implanted. As the photoluminescence experiment does not fit within the scope of this book, we will say the minimum and come back to far-infrared.

C. Possible Models of EL2 and EL2* Seen from Far-Infrared Measurement

So far we have avoided saying anything about the identity of EL2. All we have said is that EL2*, or the metastable state of EL2, is connected with a shallow donor impurity. This has been inferred from the disappearance of the donor Zeeman transition after the action of extrinsic light, or the xenon lamp light through an IR-filter. Exposure to extrinsic light leads to the photoquenching and hence to the formation of EL2*.

Our far-infrared work does not introduce a new model for the identity of EL2 itself. For the moment, accordingly, we shall simply give credit to the accepted concept that EL2 is a deep donor, possibly a double donor. On extrinsic excitation, one of the electrons bound to EL2 is freed and the center remains positively charged. The freed electron travels within the conduction band and finds a niche at an ionized donor site. Thus one can observe the Zeeman absorption of shallow donors right after the illumination even in a semi-insulating crystal. Practically all the shallow donor impurities are considered to be ionized, before illumination, on account of compensation. What we have confirmed is that the Zeeman signal thus created by the illumination gradually loses its intensity with passage of time.

We are tentatively suggesting that this is caused by association of the neutralized shallow donor with the singly ionized EL2 center. The association will require a certain voyage of the ionized defect, and the time needed for this voyage will be the time constant of photoquenching. Then there arises a question. Why does the ionization of EL2 facilitate the migration of this defect? This is not a question easy to answer. For a possible answer, we have to look for another speculation.

Generally speaking, a donor impurity cannot be considered to become mobile on ionization. Whether the donor is single or double, its ionized state has no reason to make a migration. Only under a certain condition, however, can one expect a chance for the migration. We shall describe below an idea connected with such a possibility.

It has been strongly encouraged to associate EL2 with the so-called antisite. As_{Ga}, i.e., an As atom in the substitutional site of Ga (Elliot et al., 1984; Bäumler et al., 1985). Such an antisite, a pentavalent atom replacing a trivalent atom, will certainly provide a kind of donor. It is not easy to say whether this donor is single or double. The difference in valence may suggest a double donor state. But the III-V compound crystal construction favors more covalency than ionicity. The requirement of four bonding orbitals would allow only one extra bound electron. On this ground, the defect may behave like a single donor. Recently, some modification is whispered against the identification of EL2 as a single antisite. That is to introduce a combination of $(As)_{Ga}$ antisite with another defect; 1) with a shallow impurity (Levison, 1983), 2) with a second or more than one antisite (Frank, 1985), or 3) with an interstitial As atom (von Bardeleben et al., 1985). Let us seek a situation in which an athermal ionization of such a complex by infrared light leads to a migration of some defect.

As a possibility, we propose a pair of As_{Ga} antisites. This is the next simplest model to an isolated antisite. The separation of two antisites is an unknown parameter for the moment, but can be pretty close. The pair consists of two donors and composes a kind of two-state system something like a hydrogen molecule. One may take this complex as a double donor, since it contains two electrons ready for release. One can expect both bonding and antibonding states for the pair. In thermal equilibrium, of course the system is assumed to be in the bonding state. On absorption of an infrared photon, one of the electrons will be elevated energetically, and the system may be brought to the antibonding state. Then the pair will be dissociated and one of the As atoms, being pushed into an interstitial position, may start migration and finally settle down near an ionized shallow donor impurity, sharing an electron with this donor. This model, however, does not account for the first neutralization of the shallow donor at the beginning of illumination.

As the second possibility, let us simply assume ionization of an electron from this pseudo-hydrogen molecule. The freed electron will promptly neutralize a shallow donor that has been ionized on account of compensation. Thus one can observe the Zeeman absorption of the shallow donor state. The EL2 complex after the ionization is now something like a hydrogen molecular ion, or like H_2^+. This is again a two-state system, and the shared single electron may subject to either bonding or antibonding. We may recall here that the separation between two protons is larger for H_2^+ than for H_2. In fact, it is 1.06 Å for H_2^+ and 0.741 Å for H_2. Translation of these values into the GaAs medium yields distances of ~100 Å if the effective mass theory is applied. Application of simple effective mass theory is of course dangerous, since the situation is more or less concerned with deep, and not shallow, donors. Perhaps one may have to deal with smaller separation of two As^+. In any case, one may expect that removal of an electron will tend to make the distance between two antisite atoms (or ions) larger. One of the antisites may thus have a chance to be converted to an interstitial atom (or ion). So long as this H_2^+-like complex remains stable, the neutralized shallow donor can also remain stable. But the illumination by infrared light still continues and perturbs the complex. Second ionization of the complex, or bringing the system into the antibonding state by the illumination, will definitely dissociate the pair composing the complex. A fragment of the complex already located in an interstitial site will then migrate as far as the neighborhood of the neutral shallow donor. If this fragment is As^+, the new complex $(D^+As^+)^{\cdot}$ will directly be formed. If the fragment is $As^{\cdot}$, on the other hand, we shall have $(D^+As^+)^{\cdot\cdot}$ instead. One of the electrons, however, will then drop to the As_{Ga}^+ site. In either case, a reasonable separation of D^+ and As^+ in the new complex would be of an ordser of ~100 Å. So long as one admits a mobile $As^{\cdot}$ or As^+ interstitial, an H_2^+ simulation is not so forcible a model. Since 1.06 Å is twice the Bohr radius, the distance twice the effective Bohr radius in a semiconductor GaAs in fact would be ~100 Å.

The third possibility would be a pair of As atoms, one in the antisite and the other in the interstitial site. The interstitial As atom may be mobile, but it is loosely bound to the antisite As atom on account of "molecular binding." On ionization of an As atom, either antisite or interstitial, the binding energy is decreased, so that the interstitial As or As^+ may start migration. Of course the ejected electron will have neutralized a nearby shallow donor impurity almost instantaneously. The association of As or As^+ with the neutralized donor will take some time. So a lapse of time is observed for photoquenching of the donor Zeeman absorption. The third possibility is so close to the second that detailed description will not be necessary. Indeed, it is hard to judge which process is more probable. As long as one allows for the existence of an interstitial in the migration process, however, the minute difference in the start is not very important. Probably both cases can be possible in actual processes.

Up to now, we have been dealing with the identity of EL2 on the basic idea of two As atoms, either one or both of them being antisite(s). Of course one cannot entirely forsake the possibility that a single antisite As is EL2 itself. One would then have to introduce a mechanism by which ionization of EL2 results in an interstitial which can migrate to the neighborhood of a shallow donor impurity.

In contrast to the possibility of the single antisite model, there exists a model that suggests a cluster of As_{Ga}, or $(As_{Ga})n$ with $2 \leq n \leq 5$ (Frank, 1985). This cluster model, however, is not necessarily contradictory to the pair model that we have already discussed. In fact, n can be as low as 2, while it cannot be 1. As we have already seen, the separation between two excess (antisite as well as interstitial) As atoms is expected to be ~100 Å, if we stand on the molecular approximation. They say an average concentration of EL2 in a LEC-grown crystal is $\sim 10^{16}$ cm^{-3}. That means an average separation of several hundred Å. Under such a circumstance, it would not be easy to entirely isolate a pair of two antisites from the rest. There would be several choices of combination in making a pair for each EL2 center. One might as well regard such a situation as a cluster rather than as a group of pairs of antisites. In some cases, it would be nice to treat a cluster like a group of independent pairs of antisites. In other cases, it would be better to consider a resonant jumping state of electron over more than two antisites than to consider separate pairs of antisites.

We have seen several possibilities for the model of EL2. Specifically, we are suggesting a pair of antisites or one antisite and one interstitial. This is to account for the formation of EL2*, and for the metastable state of EL2, in connection with a shallow donor impurity. One has to admit the existence of an intermediate stage, preceding the formation of EL2*, a migration process of an interstitial. Otherwise the temporal observation of the donor Zeeman absorption at the beginning of illumination cannot be explained. Our far-infrared measurement requires an association with a shallow donor impurity only after the formation of EL2*. On the contrary, or rather in a sort of similarity, there is a model by Levinson (1983), who suggests an association with the donor impurity in the ground state of EL2. In the absence of any objective observation, we are not at a position to make a comment on his model. We might, nevertheless, as well raise some speculation rather than remaining silent. If one gives credit to the H_2-like model of EL2, it can be regarded as a pair of neutral donors. If one simply takes an antisite As as the proper EL2 itself, then its association with a neutral donor follows automatically. After illumination, this "donor association" is destroyed also in our model. The final association with a shallow donor in the formation of EL2* is a matter entirely different from the Levinson model. In other words, "EL2" is associated with different donors in its ground and metastable states. This is one speculation. There might be another speculation that involves the same donor impurity. Since the average concentration of the genuine shallow donor impurities, such as Si or S, in our undoped crystals is rather low in comparison with that of EL2, it appears somewhat difficult to take all EL2 centers as accompanied by shallow donors. As already stated, the average concentration of shallow donors is 5×10^{13} cm^{-3}. That of EL2, on the other hand, is expected to be $\sim 10^{16}$ cm^{-3}. If these reasonings are justified, we are bound to suggest that the donor in Levinson's model is a neutral antisite or interstitial As atom.

In our model, we take it granted that an interstitial As or As^+ can migrate through the GaAs crystal. This assumption largely depends on the so-called Bourgoin mechanism (Bourgoin and Corbett, 1975). It has been confirmed that an interstitial As is mobile at room temperature (Frank, 1985). But we are talking about a situation at 4.2 K. The process suggested by Bourgoin and Corbett is athermal, however. If the equilibrium configuration of the defect is considerably charge-state dependent, as expected in the transition from H_2- to H_2^+-like structure, successive release and capture of an electron, following the infrared illumination, can result in a migration of the defect. In fact two possible potential curves have been suggested by Bourgoin and Corbett. One corresponds to the ground state of the defect with a bound electron and the other to the excited state: ionized and having a path with saddle points that facilitate the migration. It seems that the theory considers a migration of the defect as a whole. Apparently, however, the same theory can be applied to a migration of a fragment of the defect, an interstitial As^+ or As.

V. Concluding Remarks

In the preceding sections, we have briefly surveyed three types of imperfections in semiconductors. Crystal dislocation is a familiar imperfection, but its reaction against electron transport assumes an unexpected form in cyclotron resonance linewidth. One has to be careful when dealing with electron transport in a semiconductor containing dislocations.

The "natural quantum wells" observed in ZnSe will no doubt influence a general transport treatment for II-VI compounds. Not to speak of the twin crystal boundaries, even the existence of regular arrays of small-scale stacking faults will create an entirely new view in the practical handling of carrier mobility. The resultant two-dimensional electron system is putting forward an unexpected physical interest. Especially, such a high mobility of the two-dimensional degenerate electron gas in the zinc-dip material is quite worth a further investigation.

Our third concern is about the so-called EL2 in semi-insulating GaAs crystals. It has been quite an experience that photoquenching can be observed also in the far-infrared. The role of a shallow donor impurity comes to light in association with the metastable state of EL2. The topic itself is up-to-date and still open to various questions. The far-infrared Zeeman transition as well as cyclotron resonance will reinforce the role of electron spin resonance approach that undoubtedly supplement the photoluminescence measurement. Indeed, many things are to be done. Detection of the Zeeman transition in the relevant deep donor impurity would be just one example.

Extension of infrared and millimeter wave application to the imperfection study is boundless. If a dynamical method, such as has been described by the author in past articles in *Infrared and Millimeter Waves,* is combined for exploring the nature of defects, it would not take much time until the long-wave optics not only proven indispensable, but light the way to the next breakthrough.

Acknowledgments

The author is indebted to his collaborators, T. Ohyama, H. Nakata and K. Yamaguchi, for carrying out individual work. Zinc selenide crystals have been supplied by Professor K. Igaki and his group at Tohoku University.

The author deeply appreciates Ken Button's invitation to write a chapter for this book.

References

Bäumler, M., Kaufmann, U., and Windscheif, J. (1985). *Appl. Phys. Lett.* **46,** 781–783.

Bajaj, K. K. (1970). *Solid State Commun.* **8,** 1907–1909.

Bois, D., and Vincent, G. (1977). *J. Phys. (Paris)* **38,** L351–L353.

Bourgoin, J. C., and Corbett, J. W. (1975). *Inst. Phys. Conf. Ser.* **23,** 149–163.

Conwell, E., and Weisskopf, V. F. (1950). *Phys. Rev.* **77,** 388–390.

Cottrell, A. H. (1953). in *Dislocations and Plastic Flow in Crystals,* Clarendon Press, Oxford.

Dexter, D. L. (1952). *Phys. Rev.* **85,** 936–937.

Dexter, D. L., and Seitz, F. (1952). *Phys. Rev.* **86,** 964–965.

Elliot, K., Chen, R. T., Greenbaum, S. G., and Wagner, R. J. (1984). *Appl. Phys. Lett.* **44,** 907–909.

Erginsoy, C. (1950). *Phys. Rev.* **79,** 1013–1014.

Frank, W. (1985). *Proc. 12th International Symposium on Gallium Arsenide and Related Compounds, Karuizawa 1985,* Inst. Phys. Conf. Ser. No. 79; 217–222.

Hölscher, H. W., Nöthe, A., and Uihlein, Ch. (1985). *Phys. Rev.* **B31,** 2379–2387.

Igaki, K., and Satoh, S. (1979). *Jpn. J. Appl. Phys.* **18,** 1965–1971.

Kawamura, H., Saji, H., Fukai, M., Sekido, K., and Imai, I. (1964). *J. Phys. Soc. Jpn.* **19,** 288–296.

Landwehr, G., and Uchida, S. (1985). *Proc. Yamada Conf. XIII on Electronic Properties of Two-Dimensional Systems (EP2DS VI), Kyoto 1985,* 765–774, also in *Surface Science,* **170** (1986) 719–726.

Levinson, M. (1983). *Phys. Rev.* **B28,** 3660–3662.

Lin, A. L., Omelianovski, E., and Bube, R. H. (1976). *J. Appl. Phys.* **47,** 1852–1858.

Martin, G. M., Mitonneau, A., and Mircea, A. (1977). *Electronics Lett.* **13,** 191–193.

Matsuda, O., and Otsuka, E. (1979). *J. Phys. Chem. Solids* **40,** 809–829.

Miyake, S. J. (1965). *J. Phys. Soc. Jpn.* **20,** 412–422.

Murase, K., Enjouji, K., and Otsuka, E. (1970). *J. Phys. Soc. Jpn.* **29,** 1248–1257.

Ohyama, T. (1982). *J. Phys. Soc. Jpn.* **51,** 1431–1440.

Ohyama, T., Murase, K., and Otsuka, E. (1970). *J. Phys. Soc. Jpn.* **29,** 912–924.

Ohyama, T., Otsuka, E., Matsuda, O., Mori, Y., and Kaneko, K. (1982). *Jpn. J. Appl. Phys.* **21,** L583–L585.

Ohyama, T., and Otsuka, E. (1983). in *Infrared and Millimeter Waves,* **8** (K. J. Button, ed.), Academic Press, New York.

Ohyama, T., Otsuka, E., Yoshida, T., Isshiki, M., and Igaki, K. (1984a). *Jpn. J. Appl. Phys.* **23,** L382–L384.

Ohyama, T., Otsuka, E., Yoshida, T., Isshiki, M., and Igaki, K. (1984b). *Proc. 17th Int. Conf. Physics of Semiconductors, San Francisco 1984,* Springer-Verlag, New York, 1313–1316.

Ohyama, T., Otsuka, E., Yoshida, T., Isshiki, M., and Igaki, K. (1985). *Proc. Yamada Conference XIII on Electronic Properties of Two-Dimensional Systems (EP2DS VI), Kyoto 1985,* 123–128, also in *Surface Science 170* (1986) 491–496.

Ohyama, T., Nakata, H., and Otsuka, E. (1985). *Jpn. J. Appl. Phys.* **24,** L641–L643.

Otsuka, E., Murase, K., Iseki, J., and Ishida, S. (1944). *Phys. Rev. Lett.* **13,** 232–233.

Otsuka, E., Murase, K., and Iseki, J. (1966). *J. Phys. Soc. Jpn.* **21,** 1104–1111.

Otsuka, E., and Yamaguchi, K. (1967). *J. Phys. Soc. Jpn.* **22,** 1183–1190.

Otsuka, E., Ohyama, T., and Murase, K. (1968). *J. Phys. Soc. Jpn.* **25,** 729–739.

Otsuka, E., Fujii, Katsumasa, and Kobayashi, K. L. I. (1973). *Jpn. J. Appl. Phys.* **12,** 1600–1605.

Otsuka, E. (1981). in *Infrared and Millimeter Waves.* **3** (K. J. Button, ed.) Academic Press, New York.

Otsuka, E. (1984). *Science Report, College of General Education, Osaka University* **33,** 9–28.

Otsuka, E., Ohyama, T., and Fujii, Ken-ichi (1984). *Int. J. Infrared and Millimeter Waves* **5,** 159–169.

Otsuka, E., Ohyama, T., and Nakata, H. (1985). *Proc. 12th International Symposium on Gallium Arsenide and Related Compounds, Karuizawa 1985,* Inst. Phys. Conf. Ser. No. 79, 223–228.

Rode, D. L. (1970). *Phys. Rev.* **B2,** 4036–4044.

Tajima, M. (1985). *Jpn. J. Appl. Phys.* **23,** L690–693.

Van Royen, J., De Sitter, J., and Devreese, J. T. (1985). *Phys. Rev.* **B30,** 7154–7169.

von Bardeleben, H. J., Stievenard, D., and Bourgoin, J. C. (1985). *Appl. Phys. Lett.* **47,** 9870–972.

CHAPTER 5

SIT as Ballistic Device

Jun-ichi Nishizawa

Research Institute of Electrical Communication,
Tohoku University
Sendai 980, Japan

Kaoru Motoya

Semiconductor Research Institute, Sendai 980, Japan

I. Introduction

The three terminal solid state transistors, such as the bipolar, field effect and static induction transistors, have an inherent advantage owing to their irreversibilities between the input and the output, while in the amplifier constructed by the diodes, such as the variable capacitance diode and the negative resistance diode, the circuit needs the circulator or critical matching circuit. Therefore a new transistor which can operate at higher frequency is needed in place of diode amplifiers.

On the other hand, due to increasing need to expand the large amount of information transmission, it is necessary to raise both the carrier frequency and the bandwidth of the sideband which is introduced by the modulation signal. Therefore even in the case of optical communication, high-frequency devices are necessary for a modulation and a demodula-

ISBN 0-12-147700-2

tion not only for high frequency amplifiers. Also, the very high switching speed transistor is needed for various kinds of integrated circuits to be used in future computers, and in this point of view, the high cutoff frequency transistor is needed eagerly for communications and for high speed digital circuits.

The concept of the ballistic devices seems to have existed long ago and Bethe's "diode theory" was presented in 1942 (Bethe, 1942). The theory by Bethe derived from the assumption that the electrons can flow across the thin potential barrier without collision by the crystal lattice. This theory has not been used.

However if the condition of $D/L = \sqrt{\frac{D}{\tau}} > \sqrt{\frac{kT}{2\pi m^*}}$ is satisfied, the transport phenomenon of the device will be treated by the diode theory and this phenomenon will be seen in the Esaki diode as already pointed out by Nishizawa (1960). Since then the idea of the metal base transistor was proposed by Mead (1960) but a transistor of this type was not realized. Persky published the theory of the diffusion currents in the bipolar transistor when its base width is lower than the mean free path of carriers in 1972 (Persky, 1972).

Nishizawa et al. have shown that the diode theory will be applied when the effective thickness of the potential barrier between a paired gate is decreased and reaches the condition $D/Wg > \sqrt{kT/(2\pi m^*)}$. In the course of study of the realization and the analysis of the static induction transistor they named this transistor the ideal (or thermionic) static induction transistor (Nishizawa et al., 1975; Nishizawa and Yamamoto, 1978). Later Eastman and Shur named this transistor as the ballistic transistor in 1980 (Eastman and Shur, 1980). It is considered that when the penetrating layer thickness is thin compared to the mean free path of carrier and the amount of scattered-back carriers in the gate region towards source is neglected, then the thermionic emission equation (i.e. diode theory) will be applied for the operating mechanism of the ideal SIT (static induction transistor).

However the delay time due to the scattering process in the injection mechanism imposes a stricter limitation than that caused by the amount of scattered-back carriers in the effective gate region on the design criterion for a high-frequency cutoff transistor. Even if the detail of the scattering mechanism prevails in the carrier transport phenomenon such a situation mentioned above, it can be said that this kind of the transport phenomenon will be important for the ideal static induction transistor, for conventional transistors such as bipolar and FET and for the various kind of diodes.

II. SIT

A. Introduction to SIT

The static induction transistor (SIT) which was invented by Watanabe and Nishizawa (1950) had the multichannel structure as shown in Fig. 1, and was expected to control the flowing current through the static induction effect by the gate electrode. However, the *I-V* characteristics were not published at that time.

In 1952, Watanabe and Nishizawa published the theoretical analysis of the *I-V* characteristics of the *n-i-n* diode (Watanabe et al., 1951), and

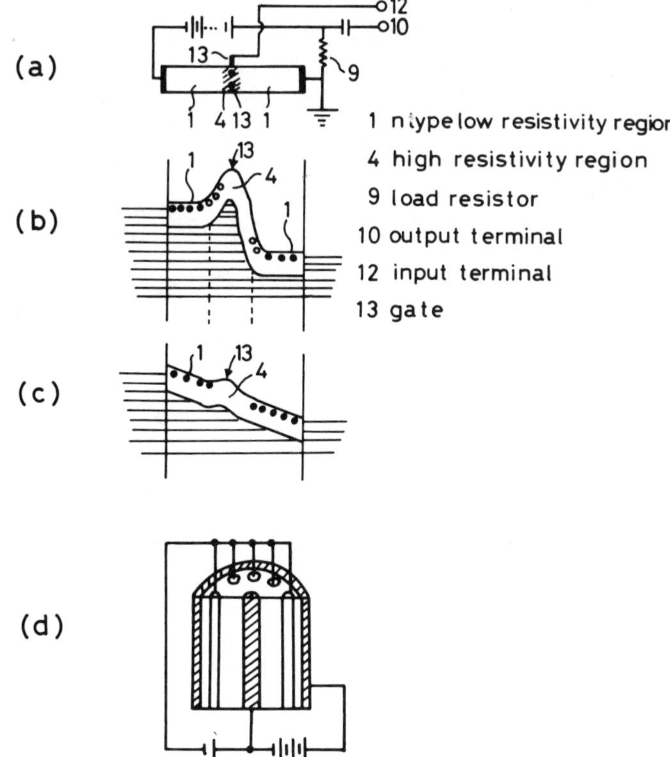

Fig. 1 Original Figures (a) through (c) of "analog-type transistor" proposed by Watanabe and Nishizawa and the analog transistor (d) by Shockley: (a) Construction of analog-type transistor and its circuit. (b) Energy band diagram of (a) when the gate is biased in the reverse direction. (c) Energy band diagram of (b) when the gate is biased in the forward direction.

concluded that it shows the space charge conduction law; $I \propto V^2$, $V^{3/2}$, and V but not triode-like characteristics.

In 1952, Shockley published the idea of the analog transistor (Shockley, 1952) and in 1953 published the same results of the space charge conduction law (Shockley and Prim, 1953; Dacey, 1953). He also concluded that the I-V characteristics of this analog transistor should follow the space charge conduction law (Shockley, 1952).

After that, many papers have been published about the analog transistor by Zuleeg (1967), Teszner (1964), Shunka (1969) and others. But in 1972, Nishizawa, Terasaki and Shibata (Nishizawa et al. 1972) published the fundamental properties of SIT in which I-V characteristics showed the variable resistance characteristics or exponential characteristics as the first experimental realization of the triode-like I-V characteristics never seen other bipolar and FETs in place of the space charge conduction law.

We have concluded in the first paper that this novel I-V characteristics in SIT is due to the reduction of the negative feedback series resistance r distributed along the channel of the field effect transistor.

This means that saturation characteristics of the field effect transistor (FET) is caused by the negative feedback of the series channel resistance. Usually it was shown $G_m \fallingdotseq 1/r_s$ and after pinch-off, the exponential characteristic was used (Nishizawa et al., 1975).

B. Basic Principle of SIT

An example of I-V characteristics of SIT is shown in Fig. 2, which gives a very beautiful exponential relationship both by gate voltage and also by the drain voltage to the drain current (Nishizawa et al., 1975). In the high current region, the effect of the series resistance appeared and the calculated value of series resistance gives reasonable value as already shown in the following paper (Mochida et al., 1978). Fig. 3 gives the I-V characteristics of SIT before pinch-off and after pinch-off, which shows the continuous variation of series resistance (Mochida et al., 1978).

The temperature dependence of SIT also changes from positive to negative as shown in Fig. 2, which is very important for high power operation, the change increases the power handling capabilities because it is effective in preventing runaway thermal burnout. It is the advantage of SIT over the commonly used bipolar transistors, in which thermal failure and permanent destruction occur due to the rise of the junction temperature.

Fig. 4 gives the results of measurement in SIT (Mochida et al., 1978). The voltage amplification constant μ is a function of the thickness of the

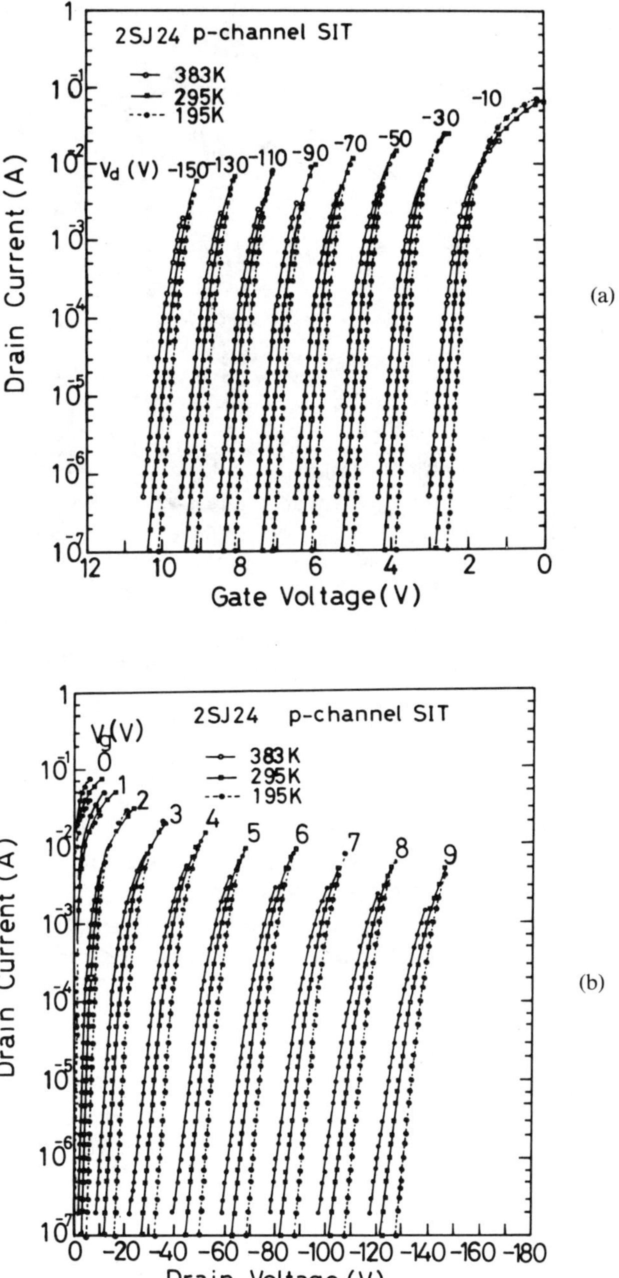

FIG. 2 Current-voltage characteristics in SIT: (a) I_d vs V_g and (b) I_d vs V_d.

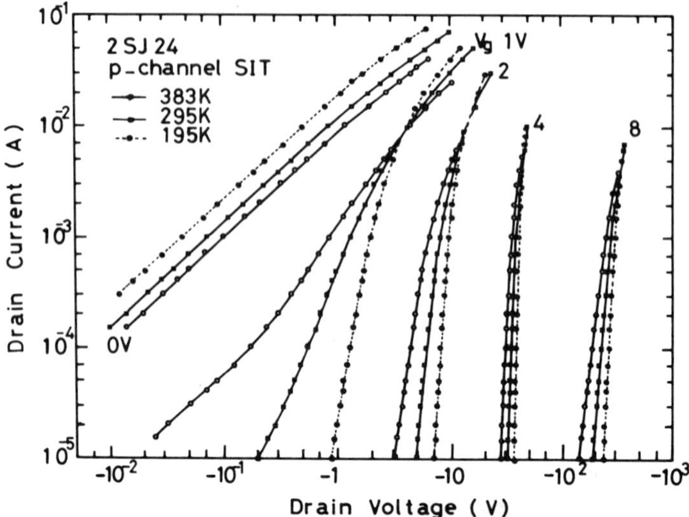

FIG. 3 Drain current versus drain voltage for three temperatures in SIT where the current transport mechanism is seen to change from the resistor-like operation to carrier injection control with increasing gate voltage.

depletion layer, W_G and W_D represent the gate toward the source and toward the drain, respectively, and

$$\mu = \frac{W_G + W_D}{W_G} = 1 + \frac{W_D}{W_G} \tag{1}$$

as published before (Nishizawa et al., 1975). Then if the regions between gate and drain are prepared not to contain any impurities, the value of W does not change and the SIT becomes constant μ. But the results shown in Fig. 4 vary over 5 orders of magnitude of the current, independently of the temperature and on the applied voltage, in spite of the very large change of the flowing current as shown in Fig. 5. If the flowing current obeys the space charge conduction law, the space charge between the gate and the drain increases the voltage, compared with the voltage without space charge, to keep the flowing current.

However, the experimental results do not show any change in the voltage amplification constant. Therefore, the effect of the space charge caused from the flowing current has been difficult to observe in SIT. It is reasonable even from the standpoint of the theoretical analysis, that the effect of the space charge should be noticed only when that of the injected carriers has become more than that of the ionized impurities and should be prominent only when the modification of the potential distribution has

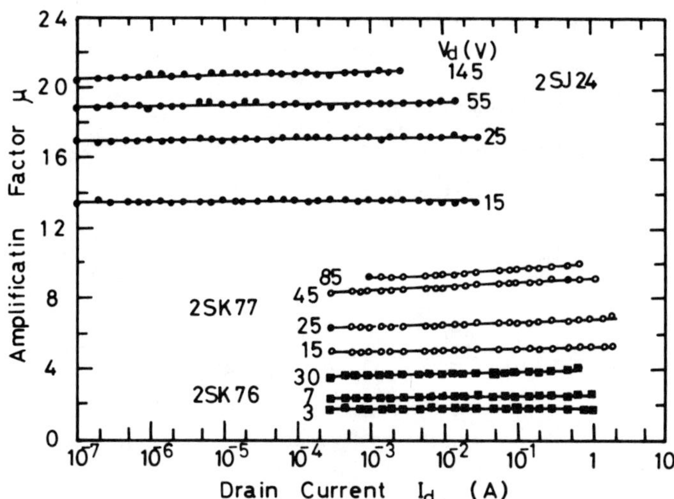

FIG. 4 Voltage amplification factor versus drain current in SIT as a parameter of drain voltage.

become more than that of the original potential distribution, which is estimated neglecting space charge caused from the flowing carriers. Usually in SIT, the applied voltage to flow the current, neglecting the space charge caused from the flowing carriers, is large enough compared with the modification by the space charge mentioned above.

The flowing drain current I_d can be described as

$$I_d = qA\left(\frac{kT}{2\pi m^*}\right)^{1/2} n_g \exp\left(-\frac{\phi - q\eta V_G - q\eta V_D\mu}{kT}\right) \tag{2}$$

or by

$$I_d = qA\frac{qD_n}{W_{G'}} n_g \exp\left(-\frac{\phi - q\eta V_G - q\eta V_D\mu}{kT}\right) \tag{3}$$

which are already published in many papers.

And when the effective thickness of the potential barrier at the pinch-off point, which is named as the intrinsic gate, $W_{G'}$ becomes thin enough to neglect the scattered back electrons in the potential barrier, Eq. (2) become effective, and in this case the α cutoff frequency is determined as a function of the distribution of the initial velocity not by the diffusion phenomena becomes very high as we named as the ideal transistor and the figure of merit is a function of Gm/C (Nishizawa, 1979c) and was estimated be 718 GHz in GaAs MISSIT (Nishizawa, 1979c). Here, q is a unit charge of an electron, k is Boltzmann's constant, T is absolute tempera-

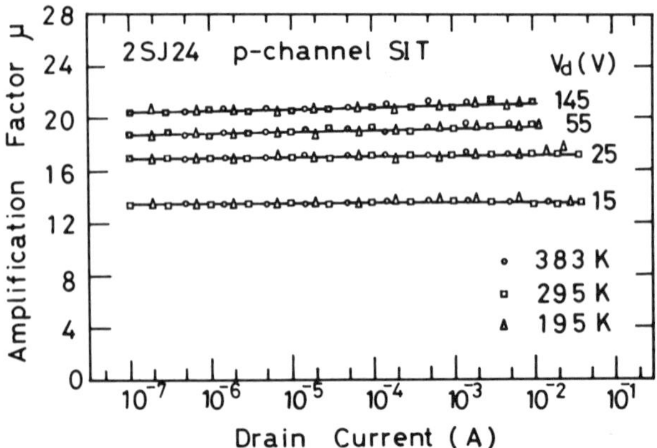

FIG. 5 Voltage amplification factor versus drain current in SIT for three different temperatures as a parameter of drain voltage.

ture, m^* is electron effective mass, n_g is electron density at the intrinsic gate, A is effective area of the channel, ϕ is diffusion potential of the intrinsic gate measured from the source, V_G is gate applied voltage, V_D is drain applied voltage, η is ratio of the voltage change at the intrinsic gate for the gate applied voltage, D_n is diffusion constant of an electron and $W_{G'}$, is the effective thickness of the potential barrier at the intrinsic gate point.

These are described as n-channel SIT and in a very early stage of our work we described an analogous equation for p-channel SIT. Also, the same relation can be obtained even for Schottky contact and MIS type SIT, because the static induction effect can be through insulator as well, mentioned even in the first papers published since 1972 (Nishizawa, 1972; Nishizawa et al., 1975). The punch-through transistor, described by Attala (1960) and Richman (1969), has been stated to show space charge conduction law.

However, MISSIT shows exponential I-V characteristics, as do the junction type, MISSIT and Schottky SIT. These have merit because of their very small amount of charge storage effect; this is very important for very fast operation.

Very recently, a permeable base transistor (Bozler and Alley, 1980) which is quite the same as buried base Schottky SIT was described. Also, a ballistic transistor (Shur and Eastman, 1980), which is quite the same as the ideal SIT (Nishizawa, 1979c) mentioned in the preceding paragraph of this chapter, is estimated to operate up to 800 GHz, which is also nearly the same as the author's estimation (Nishizawa, 1979c). Hereafter, this kind of SIT is called as the thermionic emission SIT or ballistic SIT.

C. Microwave SIT

As already has been clear, SIT can be represented to be an FET without negative feedback series resistance in the channel.

In the same time, short gate in SIT to reduce the series channel resistance Rs is also effective to reduce the capacitance between the gate electrode and the channel, C_{gch}. SIT always has a smaller time constant than that of conventional FET. Then the frequency response $\tau = r_s \times C_{gch}$ of SIT is much superior compared with those of FETs.

In the same time, normally off SIT is somewhat similar to the bipolar transistor (BPT), because the pinch-off barrier crossing through the channel is just like a base in BPT. However, the apparent base layer, which is named as virtual base, has very high base resistance and the transmission of base (gate) voltage is through the static induction effect but not through the base resistance. Therefore, SIT in this mode is like a bipolar transistor with extremely thin base. High value as a base resistance has no meaning, because SIT operates by static induction effect and even in higher frequency range, the demerit in BPT caused by the base resistance has been completely eliminated. Moreover SIT, even in the forward gate biased mode, named BSIT mode, has a much smaller storage effect and it also makes much superior performance characteristic compared with those of BPTs.

Therefore SIT can be said to be the best transistor compared with the elder transistors; BPT and FET. Morever, it has been made clear that SIT has much higher figure of merit than others, so it is possible to operate it at much smaller power level, as well as TUNNETT.

As the structure of SIT for high frequency applications, the planar type and U-SIT were proposed in place of buried gate structure as shown in Fig. 6 as the junction gate type. Planar structure can reduce the stray capacitance between source and gate. However, the important part of the vertical SIT is just at the bottom of the drain or source as shown in Fig. 6 (b), and the other part is only to keep contact with respect to the intrinsic gate point. Then the U-SIT can reduce this excess capacitance, because the excess part cut off direct capacitance between the gate electrode and the substrate has been reduced by the insertion of insulating materials with lower dielectric constant.

Fig. 7 shows the schematic structure for the MISSITs. Not only the depletion type but also the enhanced or inversion type can be realized. At the same time, it should be noticed that SIT is not only a vertical configuration but can also be a horizontal structure, as was developed in early days by Kajiwara et al. in GaAs (Kato and Kajiwara, 1977).

Negligibly small gate resistance, which can be covered by a real metal strip as shown in Fig. 6 (b) and (c), eliminates the difficulties in large area

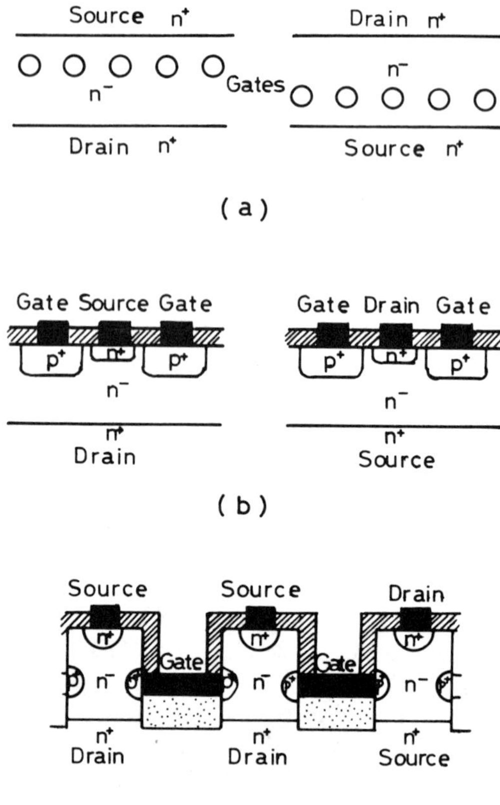

FIG. 6 Fundamental structure of *pn* junction gate SIT: (a) Buried gate type, (b) planar type and (c) U-SITs.

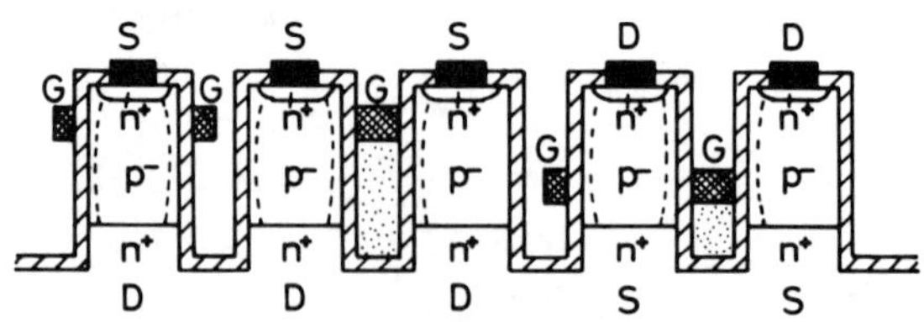

FIG. 7 Some examples of U-MIS SITs. Three in left-hand side are those in which sources are at the top; the other two are inverted structures. Some of the illustrated SITs are two-sided structures and some are single-sided structures.

devices. The limitation for the largest area comes from the preparation technology and from the matching difficulties, because the impedance of the device is inversely proportional to the area of the device. Of course in future, the propagation time delay through the contact should generate the phase mismatch of the output power, and at that moment, the idea of the distributed structure would become important. However, it is expected to be over 10 GHz. These high-power microwave SITs will replace the conventional krystron and magnetrons in the VHF to UHF frequency ranges completely owing to its high reliability and economic savings.

D. Preparation of SIT

The cross section view of the surface gate type n channel SIT is shown in Fig. 8 (a) and (b). In Fig. 8 (a) the dimension is depicted. The p^+ gate and the n^+ source are located on the surface of the n^- channel region W_{sd} in which is grown onto the n^+ substrate. The gate spacing on the surface is $2a$, in the channel its spacing is $2b$, and diffused curvature is given as r_g. In Fig. 8 (b), the actual device dimension and the source and the gate electrode are shown in case of 1 GHz Si SIT.

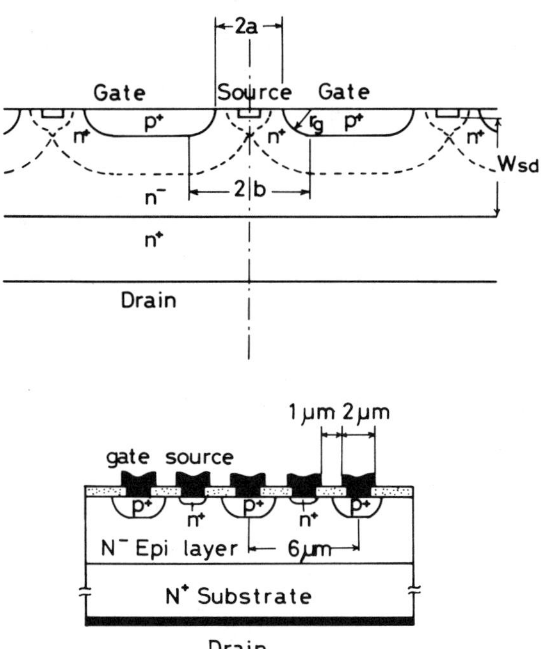

FIG. 8 Cross sectional view of planar SIT.

The fabrication steps are shown in Fig. 9. The highly resistive n^- layer with impurity concentration of $10^{14}cm^{-3}$ is grown by $SiCl_4 + H_2$ vapor phase epitaxy on the n^+ substrate with the impurity density of 5×10^{18} cm^{-3} as shown in Fig. 9 (a). Then the surface of the epitaxially grown wafer is oxdized and followed by the deposition of Si_3N_4 film from the NH_3 and SiH_4 CVD system as shown in Fig. 9 (b). After the photoresist technology as shown in Fig. 9 (c), Si_3N_4 film was removed by CF_4 plasma etching and followed by the removal of the resist. The chemical etching of S_iO_2 film was performed by the solution of buffered HF as shown in Fig. 9 (d). The p^+ gate region was formed by boron as the dopant. By the final etching process, the silicon dioxide has been also etched horizontally and the section becomes undercut as shown in Fig. 9 (e) and the dimension of side etched SiO_2 is 2 μm. The entire surface is again oxidized and its cross

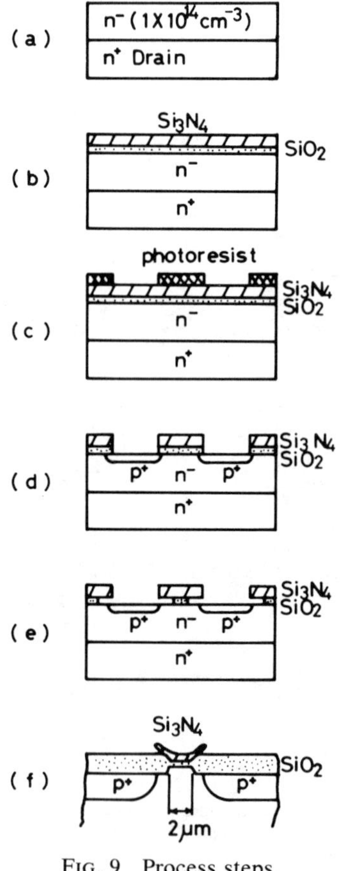

FIG. 9 Process steps.

section is shown in Fig. 9 (f). Si_3N_4 film is removed and very thin 2 μm wide oxide is left at the top of the n^- epitaxial layer without mask alignment and n^+ source region is formed after As diffusion. Metal stripe electrodes of vapor-deposited Al are aligned after selective etching of the surface oxide film. Therefore, in this case SIT is formed using only a two mask alignment process.

There are many analogous methods for the preparation of this kind of transistor. The impurity density of the gate region was diffused at least two times by this process, which automatically makes a deeper gate region and helps to pinch off the channel just at the bottom of the source region. Some examples of the *I-V* characteristics of the prepared SITs are shown in Fig. 10 (a) as for UHF and (b) as for microwave applications.

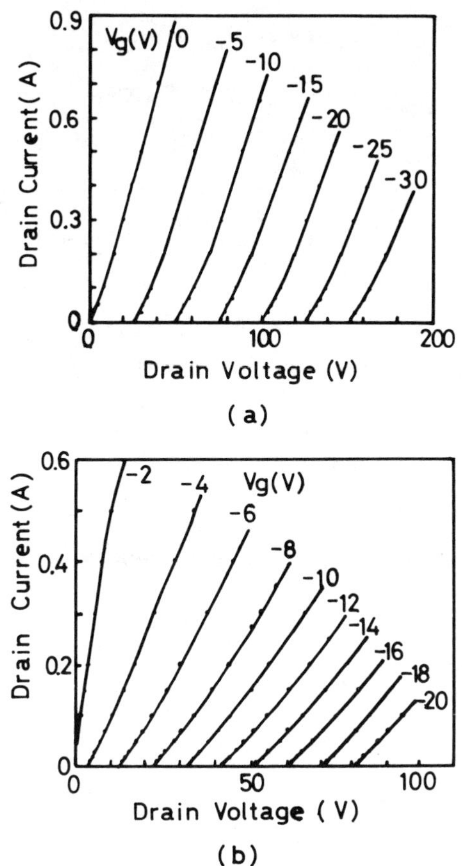

FIG. 10 Typical current-voltage characteristics of planar SIT: (a) UHF SIT and (b) microwave SIT.

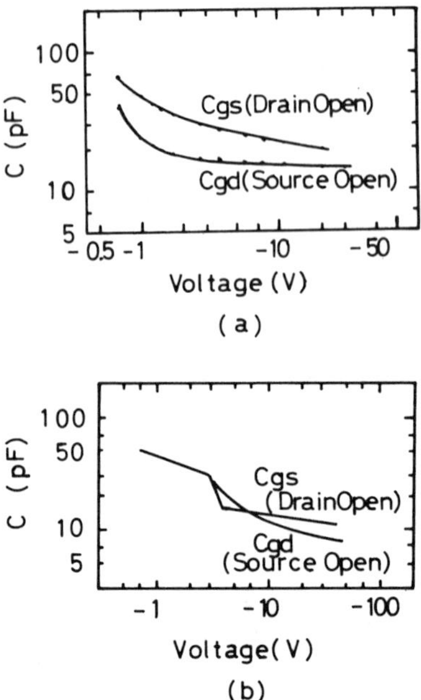

FIG. 11 Source to gate and to drain capacitance versus voltage: (a) VHF SIT and (b) microwave SIT.

Fig. 11 (a) and (b) show the corresponding voltage capacitance characteristics of Fig. 10 (a) and (b) respectively, where the small step of C_{gs} at 3 volts in Fig. 11 (b) is perhaps the evidence of pinch-off of the channel.

The imput impedance was measured to obtain the equivalent circuit of SIT. r_g, C_{gs} and C_{gd} are obtained by using the Miller effect. The input impedance is expressed as $Z = r_{gg'} + \dfrac{1}{Y_{in}}$ and $Y_{in} = g_{gs} + (|K| + 1)\, g_{gd} + j\omega\,[C_{gs} + (|K| + 1)C_{gd}]$, where $r_{gg'}$ is the gate resistance, Y_{gs} and Y_{gd} are the

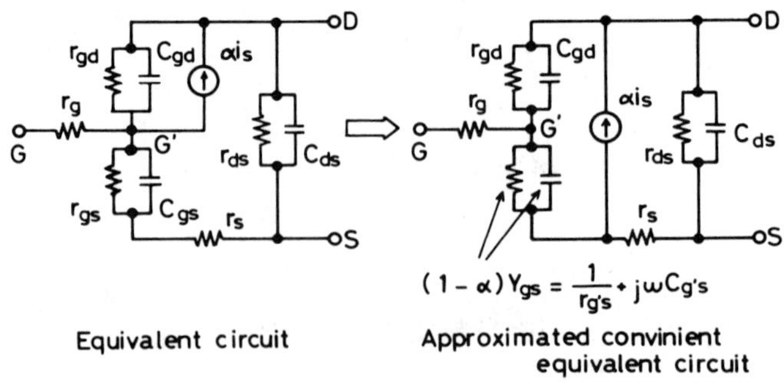

FIG. 12 Equivalent circuit of SIT.

gate-source and gate-drain admittance, g_{gs} and g_{gd} are the gate-source and gate-drain conductance. C_{gs} and C_{gd} are the gate-source and gate-drain capacitance and K is the amplification factor changed by the load resistance. Therefore changing the value of K, we can separate source-gate admittance from gate-drain admittance very easily. So the values of C_{gs} and C_{gd}, as shown in Fig. 12 measured by the only two terminal capacitance while the other electrode is open, are not so accurate.

The precisely measured parameters of the equivalent circuit of SIT are listed in Tables I and II. The high frequency characteristics of the source grounded SIT is shown in Fig. 13 (a) and (b). In Fig. 13 (a) frequency dependence of the unilateral gain and the stability factor of microwave SIT are given.

The unilateral gain of over 12dB at 100MHz and 6dB near 1GHz have been obtained. In Fig. 13 (b) the amplifier characteristics is shown, and 100W CW output power has successfully obtained at 1 GHz with the drain efficiency of about 55% where $V_{ds} = 55V$, $I_d = 3.31A$ and $V_{gs} = -10V$, respectively.

In this transistor, five transistor segments were prepared on one chip and four transistors are selected together through small inductance. The internal matching seems important to obtain resonable operations.

The idea of the potential distribution, the sectional potential distribution at the center plane of the channel from source to drain and the equivalent representation of the distributed capacitance are shown in Fig. 14. Exactly saying, three capacitances $C_{GG'}$, $C_{G'D}$ and $C_{G'S}$ should be measured to clarify the mechanism of SIT.

Aiga et al. (1980) analyzed using $T-\Delta$ transformation for three capacitances and they found that the capacitance $C_{GG'}$ resonates with the inductances distributed along the lead from gate electrode towards the ground in order to reduce the capacitance $C_{GG'}$. This resonance techniques enables the increase of the power gain as the so-called cascode connection in microwave operation.

TABLE I

TYPICAL CHARACTERISTICS OF PLANAR SIT

	VHF SIT	Microwave SIT
Gate to drain breakdown voltage	BVgd > 300 V	BVgd > 100 V
Gate to source breakdown voltage	BVgs > 100 V	BVgs > 50 V
Maximum drain current	Idmax > 2 A	Idmax > 1A
Voltage amplification factor	$\mu = 7 \sim 8$	$\mu = 5$
Out-put resistance	$r_d = 30 \sim 40\Omega$	$r_d \fallingdotseq 60\Omega$
Transconductance	Gm = 200mS	Gm = 80mS
Input capacitance	$C_{in} < 20pF$	$C_{in} < 10pF$

TABLE II

TYPICAL CHARACTERISTICS OF 1GHz 100W CW SIT

BV_{gd}	90 V
BV_{gs}	30 V
I_{dss} (at V_d 10 V)	7.2 A
G_m	1.2 S
C_{gd}	30pF
C_{gs}	63pF

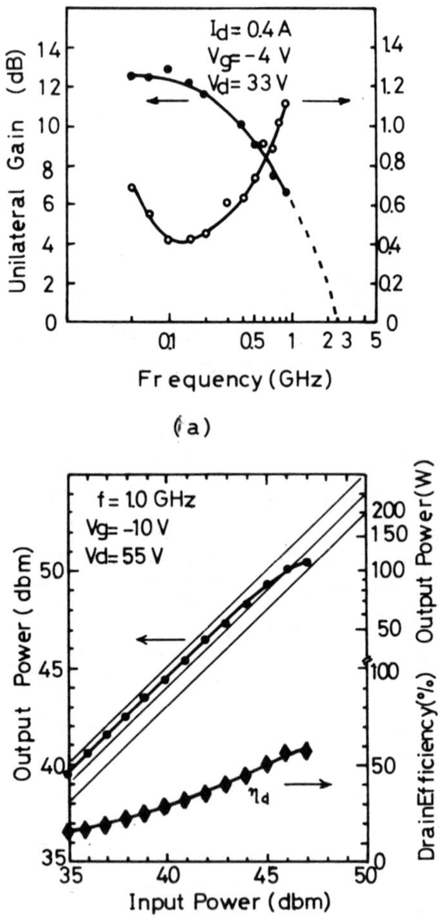

FIG. 13 (a) Frequency dependence of unilateral gain and stability factor of a microwave SIT. (b) Amplifying characteristics of 1GHz 100W SIT.

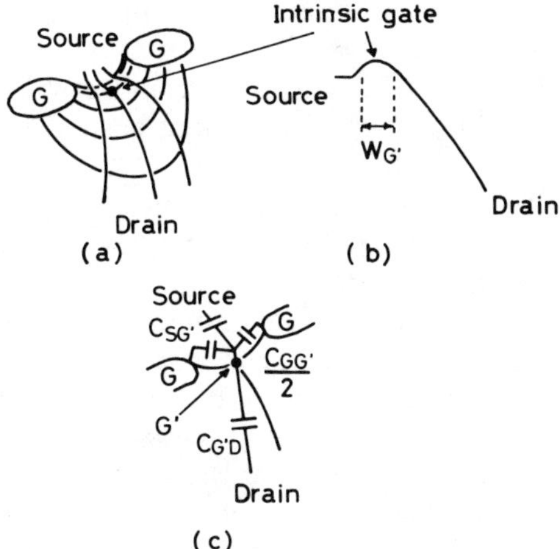

FIG. 14 Static induction effect in SIT: (a) potential profile, (b) potential profile from source to drain along channel center and (c) intrinsic capacitance.

However the real stray capacitance between the source and the gate and the drain should be separated from these *T*-connected capacitance to know the real physical phenomenon. Finally at least 6 capacitances should be analyzed. These are future problems.

The structure of planar surface gate type SIT is not the best one, since it still contains the excess stray capacitance. The step gate electrode (*U*-gate) structure as shown in Fig. 6 or 7 should be introduced. In addition, the internal matching of input and output ports will be able to increase f_c and output power. The most impressive results are that the gate resistance can be nearly equal to zero and the breakdown voltage of the source to drain is limited by the operating frequency, since the thickness of the drain depletion layer is limited by the transit time effect of the drift carriers in a channel. If there is not any limitation for the frequency, the drain breakdown voltage can be prepared as high as several thousand volts.

The 2 GHz, 10W SIT has been developed by Toshiba's group in 1979 with use of the planar structure (Shino et al., 1979). Recently, the 6 GHz SIT with *U*-gate structure, which was fabricated by using plasma etching technique, has been presented by GTE's group (Cogan et al. 1983).

III. Ideal SIT

A. Basic Principle of Ideal SIT

As described in Section I. B, when the device dimension of SIT is less than the mean free path of an electron, SIT's f_α and f_T will be very high.

When the effective thickness of the gate region potential barrier, W_g is decreased and W_g is close to the mean free path of the carrier, the non-scattered carriers by the lattice vibration travel to the drain region over the potential barrier induced by the applied gate bias. If there is some scattering, the velocity of the carrier will be changed by the elastic collision. The probability of scattered back carriers which have already travelled from the source over the potential barrier again is quite low. Therefore nearly all of the carriers which have started from the vicinity of the source toward the drain will travel to the drain even with a certain amount of scattering by the lattice vibration. These situations are depicted in Fig. 15 (a) as the scattering case without scatter-back and Fig. 15 (b) as the non-scattering case.

Assuming that all of the carriers over the potential barrier travel to the drain, then the current-voltage characteristics of the device will be expressed by using the emission of thermal electron as follows (Nishizawa et al., 1975; Nishizawa and Yamamoto, 1978)

$$J = q\sqrt{\frac{kT}{2\pi m^*}}\, n_s \exp\left[-\frac{q\,(\phi_{Gs} - \eta V_G)}{kT}\right] \tag{4}$$

In this case, the speed of the carrier will not saturate over saturation velocity since the scattering in a crystal is quite small. The Bethe's theory called as the diode theory treated the almost non-scatter case (Bethe, 1942) and similar work has done by Persky (1972). We postulated the SIT

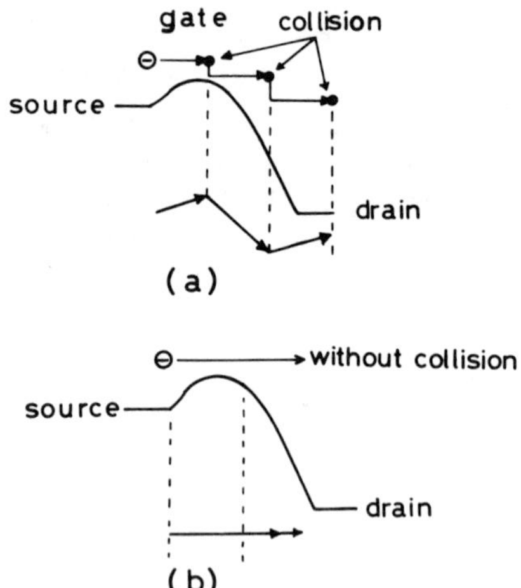

FIG. 15 Electron transport in SIT: (a) scattering case and (b) non-scattering case (ideal SIT).

in equation (4), called the "thermionic emission type" or the "ideal type" of SIT, in 1974. The estimated f_T was extremely high. The frequency limit of 780 GHz with 0.1 μm gate width in GaAs which was caused by the parastic capacitance following equation Gm/C was obtained by Nishizawa (1979c). The same result has been followed by Eastman et al. and their estimation is 800 GHz (Shur and Eastman, 1980). They named this ideal SIT as the "ballistic transistor" at that time.

The basic principle such as f_α, and f_T of the ballistic static induction transistor, differs from the conventional transistor in the carrier velocity at first. Next, f_α can be determined by the initial velocity distribution of the injected carriers from source to gate region when the collision and the scattering between the carriers and the lattice vibration are assumed to be neglected. If the initial velocity of an electron is v_s and the potential barrier height at the gate region measured for the injection point is given as $V(x)$, then the phase delay of an electron will be expressed as follows:

$$2\pi j \int_0^\infty f(v_s) \int_0^{W_D} \left[\frac{2}{m^*}\left(\frac{1}{2}\, v_s m^* - V\right)\right]^{-1/2} d\chi dv_s \tag{5}$$

where W_D is distance between the injection point at the source and the drain and $f(v_s)$ is the velocity distribution function of an electron with v_s. Exactly speaking $\left[\frac{2}{m^*}\left(\frac{1}{2}\, v_s m^* - V\right)\right]^{-1/2}$ should be the velocity of an electron at a constant energy. The variation of this integral corresponds to the current amplification factor α and the frequency which corresponds to one radian of phase angle of the above integral is called as f_T. f_α is the frequency when α reaches $e^{-1/2}$. In ordinary conditions this value is very high, so the frequency limit of the device will be determined from the stray capacitance and the mutual conductance G_m. The cutoff frequency f_c is given as follows. If the gate voltage induces charge Q at the top of the barrier, transfer conductance G_m can be expressed as follows:

$$G_m \simeq \frac{i}{V_G} = \frac{\dfrac{Q}{W_{G'}}\sqrt{\dfrac{kT}{2\pi m^*}}}{V_G} \tag{6}$$

Q is expressed by the gate voltage V_G and induced charge Q as follows:

$$Q = V_G \times C_G \tag{7}$$

where C_G is the equivalent total capacitance measured from the gate. Subscription of Eq. (7) to Eq. (6), f_c is expressed by the next equation:

$$\omega_c = \frac{G_m}{C} \simeq \frac{G_m}{C_G} = \frac{1}{W_{G'}}\sqrt{\frac{kT}{2\pi m^*}} \tag{8}$$

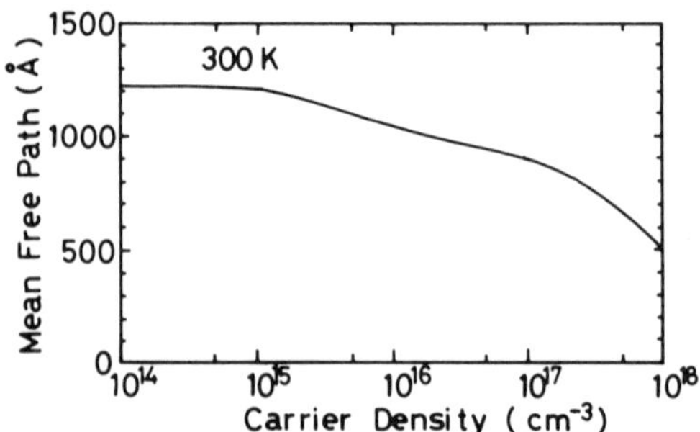

FIG. 16 Mean free path of an electron in GaAs versus carrier density.

Assuming GaAs as a material and $W_{G'} = 0.1\ \mu\text{m}$, f_c is estimated to be 718 GHz. The short gate region and low effective mass as a carrier are needed to raise the cutoff frequency of the ballistic SIT.

The mean free path of an electron is given as follows:

$$\lambda = \tau v_{th} \tag{9}$$

If λ is expressed as

$$\lambda = \frac{\mu}{q}(3kTm^*)^{1/2} \tag{10}$$

then the mean free path of an electron is calculated as a parameter of the doping density, the temperature, and the mobility of an electron, respectively. The calculated λ as a parameter of carrier density is shown in Fig. 16. When $N = 10^{16}\text{cm}^{-3}$ and 10^{18}cm^{-3}, λ is 1040Å and 510Å, respectively. The effective gate region should be thinned smaller than λ or at least nearly equal to λ. The total width of the ballistic SIT is determined by adding the transit time effect in a channel (Nishizawa and Watanabe, 1958a).

The carriers over the intrinsic gate drift in the channel and then arrive at the drain. At higher frequency, the phase rotation of the injected carriers in the channel should be considered in the design of the ballistic SIT.

Ignoring the diffusion effect, the electron current density has been given as follows (Okamoto and Nishizawa, 1966):

$$|j'| = \frac{qnv_n^2}{2\sqrt{2}\omega L}\left(1 - \cos\frac{\omega L}{v_n}\right)^{1/2} \tag{11}$$

where L is the length from the intrinsic gate to the drain, v_n is the saturation velocity of an electron as shown in Fig. 17.

However, this result is for the case of the usual drift and diffusion equation, and an electron velocity in real drift space is much higher than

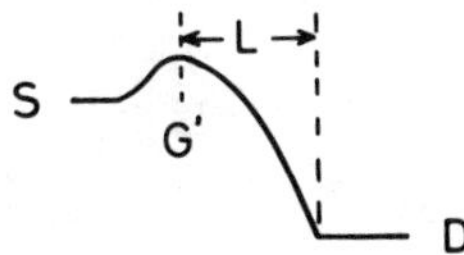

FIG. 17 Potential profile from source to drain.

this value when the thickness is within the length of the mean free path caused from the lattice scattering so that this equation gives the lower value in the actual not ballistic nor quasi-ballistic case.

If the electron has not interacted with the lattice of crystal, the velocity will be replaced by the ballistic velocity.

Even if the acceleration of an electron by the electric field was neglected, the estimated f_c value gives extremely high value. Using the average velocity $\bar{v}_n$ and the transit angle $\theta \left(\equiv \omega \dfrac{L}{\bar{v}_n}\right)$, Eq. (11) is reduced to

$$|j'| = \frac{qn\,\bar{v}_n}{2\sqrt{2}} \frac{(1 - \cos\theta)^{1/2}}{\theta} \tag{12}$$

Fig. 18 shows the calculated results of $|j/j_0|$ as a parameter of frequency. When $L = 100$nm and $\bar{v}_n = 2 \times 10^7$cm/sec, the f_c is 850GHz. Over 850GHz, $|j'|$ decreases, then increase from 2000GHz and has a second peak at 3000GHz. These phase rotation occurs with the transit angle. Decreasing the thickness to 100Å and 10Å, f_c is 8.5THz and 85THz, respectively.

In the normal condition, the transistor operates from the low frequency to the f_c region. However, the transit time effect can be used as the transit

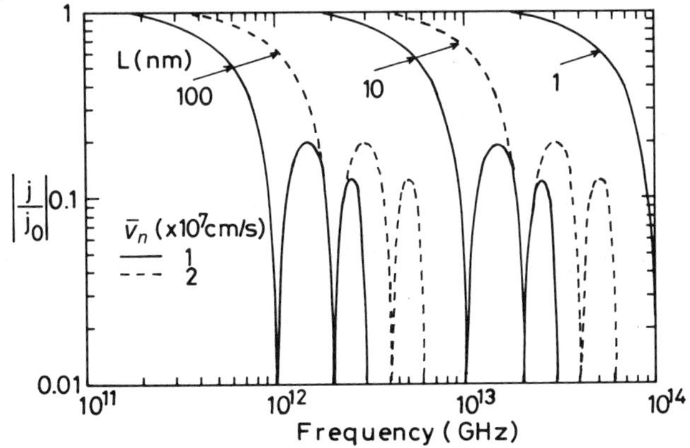

FIG. 18 $|j/j_o|$ versus frequency as a parameter of transit region width L.

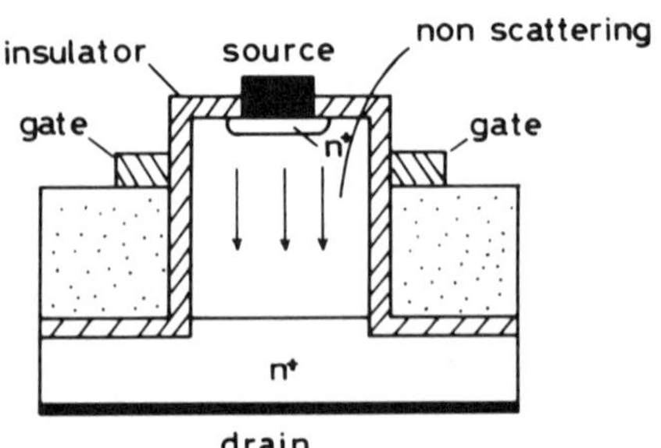

FIG. 19 Cross sectional view of ballistic SIT.

time transistor. This transistor should be called the ballistic static induction transit time transistor.

Several structures of ballistic SIT have been considered and one example is shown in Fig. 19. As the gate structure, *pn* junction gate, Schottky barrier gate, MIS (or MOS) gate and heterobarrier gate will be available for the ballistic SIT. The MIS gate will be good since the storage effect in the gate region is not seen. As described above the dimension of the device is less than about 0.1 μm at f_c around near 800 GHz, so the critical control of the impurity profile and the device fabrication technology are needed.

The perfect crystal or expitaxial layer which has no lattice defects and dislocations should be realized, since the operation principle of the ballistic SIT is based on the non-scattering process. If there are some crystal imperfections, the carriers would be scattered at that point and the velocity of the carriers would be decreased. Then the f_c of the transistor would be decreased.

There are several epitaxial growth methods for GaAs such as liquid phase epitaxy (LPE), vapor phase epitaxy (VPE), plasma CVD (PCVD) and molecular beam epitaxy (MBE). Among these methods, the epitaxial layer obtained by LPE is most excellent, but the controllability of the layer thickness of LPE is not good. The evaporation method which includes MBE has drawbacks with its crystal defects since the growth temperature is high. This situation is the same for VPE and there were defects induced by the ion species in PCVD.

We have adopted the atomic layer epitaxy produced by Suntola (1981, 1984) to overcome the current epitaxy method. Suntola et al. utilized this method to realize polycrystalline ZnSe and ZnS. We have successfully obtained the GaAs single crystal in 1984 (Nishizawa et al. 1984; Nishizawa and Kokubun, 1984).

After the set of the GaAs substrate the growth chambers are evacuated to a very high vacuum. The Ga source gas such as TEG (triethyl gallium)

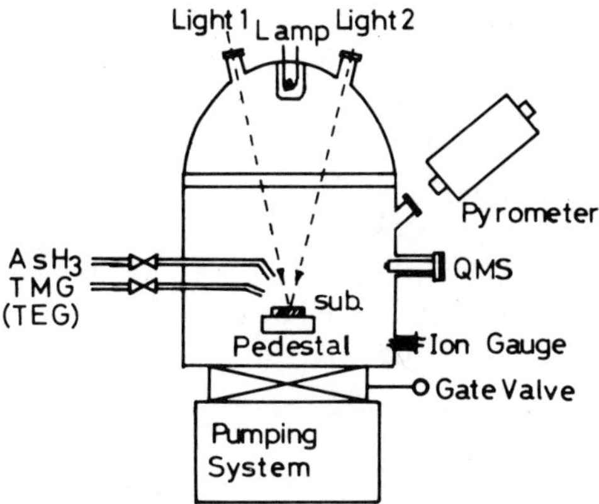

FIG. 20 Schematic illustration of the growth apparatus of the photoexcited molecular layer epitaxy (PMLE).

or TMG (trimethyl gallium) is introduced onto the GaAs surface in a certain pressure with short period. The excess TEG is evacuated. AsH_3 as a As source is introduced onto the GaAs substrate in which surface the TEG or TMG is absorbed. Then the chemical reaction has occured to produce one molecular layer of GaAs by the two different kind of gas sources. The desired repetition of this growth cycle enables one to obtain atomic accuracy in very thin layers of GaAs. We have combined this growth method and the photoexcitation process called photoexcited molecular layer epitaxy (PMLE). The ultraviolet light source such as the mercury lamp or excimer laser illuminates the surrounds of the GaAs substrate through the window as shown in Fig. 20.

The high quality GaAs epitaxial layer was successfully obtained at 290°C (Nishizawa, 1985). This temperature can be said to be the lowest temperature ever for the GaAs epitaxy and has never been obtained by another method. The detailed growth mechanism is under investigation now. There will be several wavelengths of light effective to enhance the growth rate for among the resolution of gases, the reaction for the surface absorption during the epitaxial process (Nishizawa et al., 1984). The choice of the magnitude of each wavelength of incident light may change the growth parameters. The $Ga_xAl_{1-x}As$ crystal will be obtained by adding the gas which includes aluminum. If this technique is developed more, the heteroepitaxy of GaAs and $Ga_xAl_{1-x}As$ will be obtained.

In summary, this newly developed PMLE has advantages over the conventional epitaxy method as follows:

1) The accurate monolayer is obtained by one cycle of the introduction of component gases.

2) The low growth temperature will eliminate the lattice imperfections and also helps to make a sharp impurity profile.

B. Multistage and Travelling Wave Ideal SIT

The multistage amplifier is constructed when we want to have the high gain. The connection of the two stage MIS (or MOS) gate type ballistic SIT is shown in Fig. 21 with the capacitances between electrodes.

The gate capacitance is expressed as $C_{gs} + C_{gd}(1 + |K|)$ by the Miller's effect and also the drain of the first stage has the distributed capacitance as $C_{ds} + C_{gd}\left(1 + \frac{1}{|K|}\right)$. These two capacitances are distributed in the two stage transistors. The load resistance is R_L and the capacitive impedance is equal to R_L at ωc.

Then the next equation is given as follows.

$$R_L = \frac{1}{\omega_c\left[C_{gs} + C_{ds} + C_{gd}\left(2 + |K| + \frac{1}{|K|}\right)\right]} \tag{13}$$

The gain bandwidth product $G \times B$ is expressed as

$$G \times B = \omega_c\, G_m \times Z_{in} \times |K_0| \tag{14}$$

which corresponds to the bandwidth when the amplification factor equals to unity and then $K = K_0/\sqrt{2}$ (3dB point). So Eq. (14) is reduced to

$$\begin{aligned} G \times B &\fallingdotseq \frac{1}{R_L\,(C_{gs} + C_{ds} + 4C_{gd})} \times \frac{R_L \times G_m}{2}\, G_m \times Z_{in} \\ &= \frac{G_m{}^2 R_L}{\sqrt{2}}\,[C_{gs} + C_{gd}\,(1 + \sqrt{2})] \\ &\fallingdotseq \frac{G_m}{1.4\,(C_{gs} + 2C_{gd})} \end{aligned} \tag{15}$$

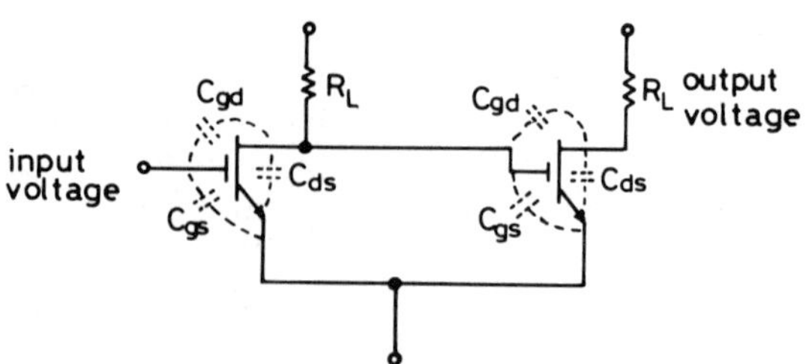

Fig. 21 Two stage *n* MIS SIT amplifier. The distributed capacitances among electrodes are shown.

where

$$Z_{\text{in}} = \frac{1}{\omega_c\,[C_{gs} + C_{gd}\,(1 + \sqrt{2})} \tag{16}$$

Then the cut off frequency is given as follows:

$$f_c = \frac{\sqrt{\dfrac{kT}{2\pi m^*}}}{8.8 W_b} \tag{17}$$

This analysis predicts that the imput capacitance of the two stage ballistic amplifier is roughly assumed to be $C_{gs} + 2C_{gd}$ in place of C_{gs}.

The multichannel structure of the ballistic SIT will be introduced for the high-power operation. In this case the capacitance will be increased in proportion to the number of the channel. As discussed in the microwave SIT, these capacitances will be decreased by the neutralization technique. Moreover, the introduction of a travelling wave structure for the ballistic SIT is more favorable. The difficulty of construct of the ramp element circuit and the easiness of the patterning in a semiconductor ballistic SIT chip will enhance the introduction of the travelling wave structure above the millimeter wave region.

The basic circuit of the distributed ballistic SIT is shown in Fig. 22 (Nishizawa et al., 1975) and its equivalent circuit is also shown in Fig. 23.

The characteristic impedance of the line is expressed as the next formula

$$Z_0 = \sqrt{\frac{R + j\omega L}{G + j\omega C}} \tag{18}$$

As shown in Fig. 23, characteristic impedance of the input line and output line, the impedance of the terminations of the output line, the termination

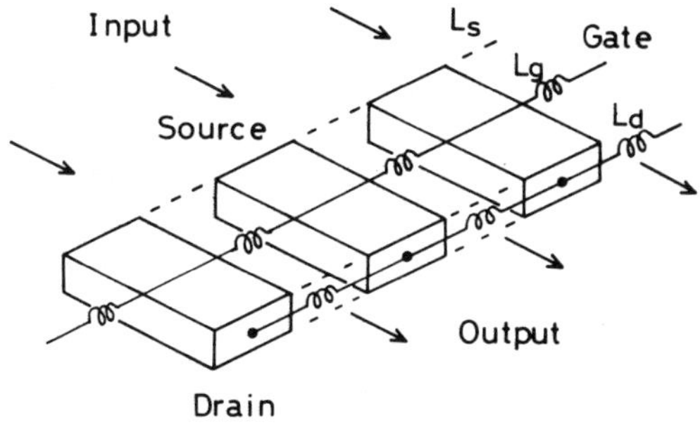

FIG. 22 Travelling wave ballistic SIT.

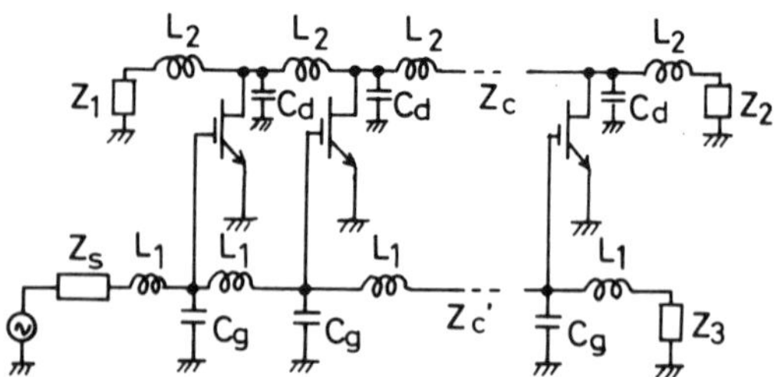

FIG. 23 Equivalent circuit of the travelling wave ballistic SIT.

of the input line and the input impedance of the input signal source are given by $Z_{c'}$, Z_c, Z_1, Z_2, Z_3 and Z_s, respectively.

Three cases were considered and described below (Kitsuregawa and Nishizawa, 1973).

Case 1. $Z_1 = Z_2 = Z_{c'}$, $Z_3 = Z_c = Z_s$

In this case perfect matching is performed, so the amplification factor will be increased in proportion to the line length as similar to the conventional distributed amplifier composed by such the discrete transistor and the ramped distributed element of L and C.

Case 2. Z_3 is open ($= \infty$) and $Z_1 = Z_2 = Z_c$

In this case there is a standing wave in the input line. The amplification factor is varied by the length of the input line and the amplification factor is higher than that of the complete matching in case 1 at a certain input line length.

Case 3. Z_1 and Z_3 are open and $Z_2 = Z_c$

There exist standing waves in both the input and output line. The amplification factor is higher than those of cases 1 and 2 at a certain length of both the input and the output line.

The open and proper impedance matching for the input and output of the ballistic SIT will be easily accomplished by the external circuit elements such as the tunable stub, the strip line or the *E-H* tuners. Generally speaking, the travelling wave ballistic SIT using case 1 circuit will be suitable for the wide band amplifier and the cases 2 and 3 will be suitable for the narrow band or single frequency amplifier, respectively. These conditions of the travelling wave ballistic SIT is depicted in Fig. 24.

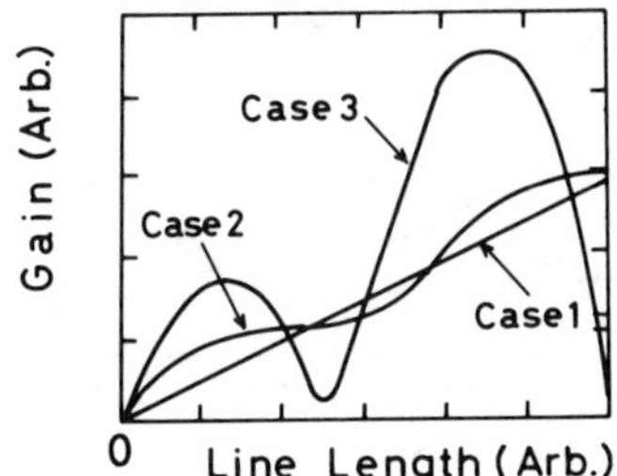

FIG. 24 Gain versus line length of the distributed line of cases 1 to 3.

C. NOISE PROPERTIES OF THE IDEAL SIT

The noise figure of every kind of semiconductor device is an important factor for the applications. It is well known that the noise figure of the GaAs FET is smaller than that of the Si bipolar transistor by virtue of the majority carrier device. The noise source from the source electrode for both FET and SIT is evaluated to be the same, and the difference between FET and SIT comes from the series channel resistance. SIT, which has smaller channel resistance r_s, yields the smaller noise level than that of the conventional FET. This is true by the fact that the Si SIT logic IC which include the I^2L (Nishizawa and Wilamowski, 1976) (Nishizawa and Wilamowski, 1977), CML (Nishizawa et al., 1979), MOS (Nakamura et al., 1978) and CMOS (Suzuki, 1983) have been operated at a very small input level. Recently developed Si photo SIT and its integrated image sensor have the small noise properties with very high optical gain (Nishizawa et al., 1979b, 1983). The low noise characteristic of the already developed SIT family transistor will be adopted for the ballistic SIT naturally.

D. ISIT IC

The ballistic SIT is applicable to both the discrete and the integrated circuits as similar to SIT and other transistors.

One of the example of the ballistic SIT inverter is shown in Fig. 25. The substrate is the semi-insulating GaAs, so the metallization of each source, gate and drain electrode is performed in the planar geometry. This IC is suitable for analogue and digital applications, so the various kind of functions such as memory, A/D and D/A conversion, and the numerical calculation circuit will be realized.

The IC using the high cutoff frequency of the ballistic SIT can be said to have promising high speed and low power consumption characteristics.

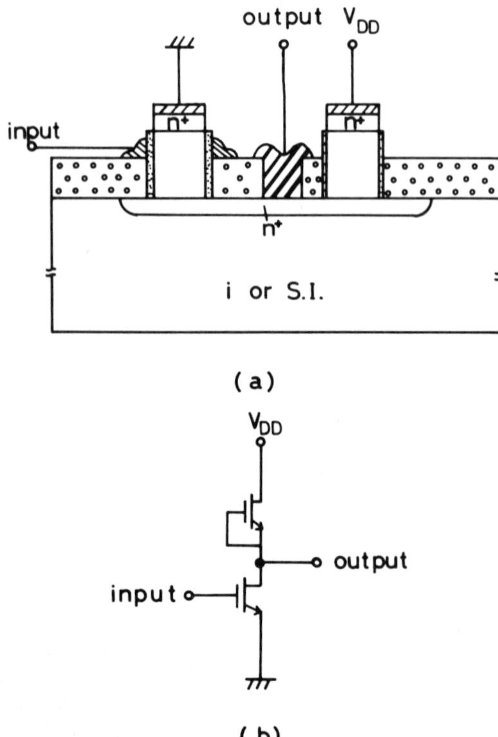

FIG. 25 Example of ballistic SIT IC: (a) cross sectional view and (b) its equivalent circuit of the inverter.

IV. Ideal SITT

A. Basic Principle of Ideal SITT

The distribution capacitance between the source electrode and the gate (or intrinsic gate) region of the ballistic transistor is quite high owing to the accumulation effect. The static induction tunnel transistor (SITT) which uses the tunnel injection as the source was proposed by Nishizawa to overcome the drawbacks of the ballistic SIT (Nishizawa, 1980). The author had already pointed out that the tunnelling current in the Esaki diode (Esaki 1958) seems to be followed by the Bethe's theory (Bethe, 1942). The forward *I-V* characteristics of the Esaki diode are restricted within the bandgap of the semiconductor p^+n^+ junction. The transit time negative resistance diode was invented by Nishizawa and Watanabe (1953) and Shockley (1954) independently.

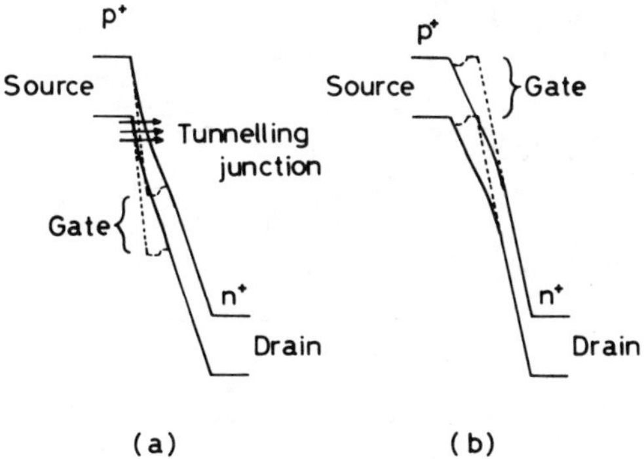

FIG. 26 Potential profile in the ideal static induction tunnel transistor (ISITT).

However, in the reverse bias tunnel junction the operating region is wide compared to the foward biased p^+-n^+ junction. The tunnel injection transit time diode, called TUNNETT, had been proposed (Nishizawa and Watanabe, 1958b) as the theoretical analysis of the avalanching negative resistance diode which is called the IMPATT diode nowadays.

The operating principle of the static induction tunnel transistor is shown in Fig. 26. The device structure is $p^+-n^+-i(n^-$ or $n)-n^+$ one which was proposed by Read (1958) and Nishizawa independently (Nishizawa and Watanabe, 1953). The active region of the diode should be depleted, then the applied gate potential controls the tunnel injection. The one cross section of one SITT is shown in Fig. 27. The equivalent circuit of the reverse biased tunnel junction is expressed as the parallel circuit of the tunnel conductance g and the capacitance $\left(=\varepsilon\frac{S}{W_t}\right)$ where ε is the permittivity of the semiconductor, S is the area of the junction and W_t is the width between the source and gate electrode, respectively.

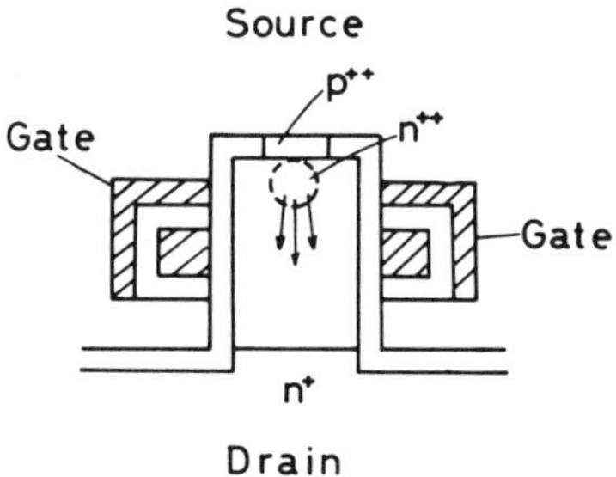

FIG. 27 Typical cross sectional view of the ideal static induction tunnel transistor.

The transconductance G_m is the ratio of the tunnel current induced by the small change of V_G as ΔV_G and it is well known that this value is higher than that of the conventional *pn* junction. This value in the reverse biased tunnel junction is larger than that of the Esaki diode. The structure of this SITT is the triode type of TUNNETT diode (Okabe et al., 1968) and the oscillation frequency has been verified up to 338 GHz by our experimental result (Nishizawa, 1982). This $f_{\max_{osc}}$ is higher than that of the GaAs FET. The $f_{\max}$ of SITT is calculated below.

The f_c of SITT is given as follows:

$$f_c = \frac{G_m}{2\pi C_g} = \frac{\dfrac{i}{V_g}}{2\pi C_g} = \frac{\dfrac{Q}{\tau}}{2\pi C_g V_g} \tag{19}$$

and

$$Q = C_g V_g \tag{20}$$

Substitution of Eq. (20) in Eq. (19) yields the next formula as

$$f_c = \frac{1}{2\pi\tau} \tag{21}$$

where τ is the tunneling transition time. The tunneling transition time τ is expressed as (Zener, 1934)

$$\tau = \frac{\hbar}{qEa} \tag{22}$$

where $\hbar$ is the Planck's constant devided by $2\,\pi$ (1.0546×10^{-34} J·sec), E is the electric field intensity of the tunnel junction and a is the lattice constant of the semiconductor material, respectively. Substitution of Eq. (22) in Eq. (21) gives f_c as follows:

$$f_c = \frac{qEa}{2\pi\hbar} \tag{23}$$

The calculated f_c as a parameter of E is plotted in Fig. 28 assuming $a =$ 5.6533Å as taken from GaAs. With increase of the electric field intensity, f_c is increased. When the electric field intensity is about 7MV/cm, f_c is estimated to 10^{14} Hz (100 THz). Such a high electric field can be realized in the heavily doped p^+-n^+ junctions. If we assume the abrupt junction, the depletion layer width and the electric field profile (triangle type) are obtained as follows:

$$W = \sqrt{\frac{2\varepsilon_0\varepsilon_s\,(V_a + V_{bi})}{q}\left(\frac{1}{N_A} + \frac{1}{N_D}\right)} \tag{24}$$

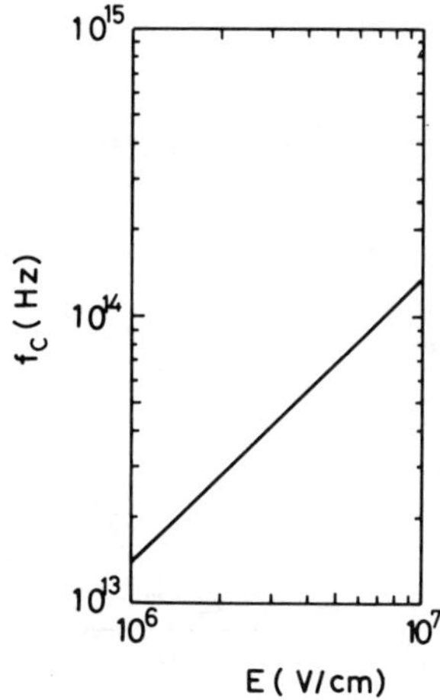

FIG. 28 f_c versus electric field intensity.

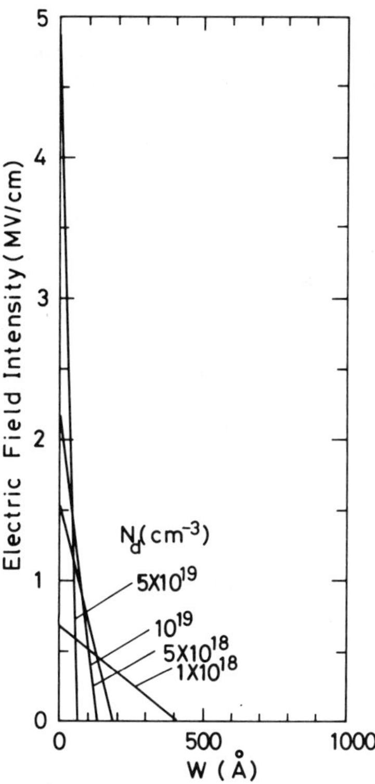

FIG. 29 Electric field profile of the abrupt p^+-n^+ junction at 0 bias.

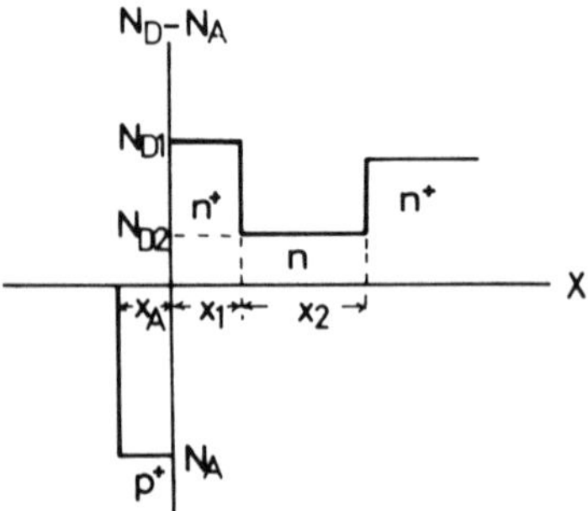

FIG. 30 Doping profile in the channel of the p^+-n^+-n-n^+ structure of ideal SITT.

$$E_{\max} = \frac{2(V_a + V_{\mathrm{bi}})}{W} \tag{24a}$$

where ε_s is the dielectric constant of semiconductor material, V_a is the applied voltage across the p^+-n^+ junction, V_{bi} is the built-in potential, N_A is the acceptor concentration of the p^+ layer. N_D is the donor concentration of the n^+ layer.

The electric field profile at zero bias as a parameter of N_D is shown in Fig. 29 where N_A is fixed to $10^{20}\mathrm{cm}^{-3}$ and material parameter is taken from GaAs. When N_D is over $2 \times 10^{18}\mathrm{cm}^{-3}$, the $E_{\max}$ exceeds 1MV/cm. The δ function like very sharp n^+ layer is needed to establish the high electric field for the tunnel injection.

The doping profile of the p^+-n^+-n-n^+ structure is shown in Fig. 30. The relation between the depletion layer width W and doping density is expressed by using Eq. (24) as follows:

$$W = \sqrt{\frac{2\varepsilon_0\varepsilon_s\,(V_a + V_{bi})}{q}\left(\frac{1}{N_A} + \frac{1}{N_{D1}}\right)} \tag{24b}$$

When N_A is fixed and $V_a = 0V$, then the depletion layer width W have been obtained. If the total active region is depleted at 0 bias voltage, x_1 should be thinned less than W. In this case the voltage needed to deplete n layer is obtained by next equation as follows:

$$V = \frac{qx_2^2}{2\varepsilon_0\varepsilon_s} N_{D2} \tag{25}$$

or

$$N_{D2} = \frac{2\varepsilon_0\varepsilon_s}{qx_2^2} V \tag{25a}$$

If 90% of the built-in potential is used to deplete the p^+n^+ region, then 10% of built-in potential can be used to deplete the n layer. If V and x_2 are

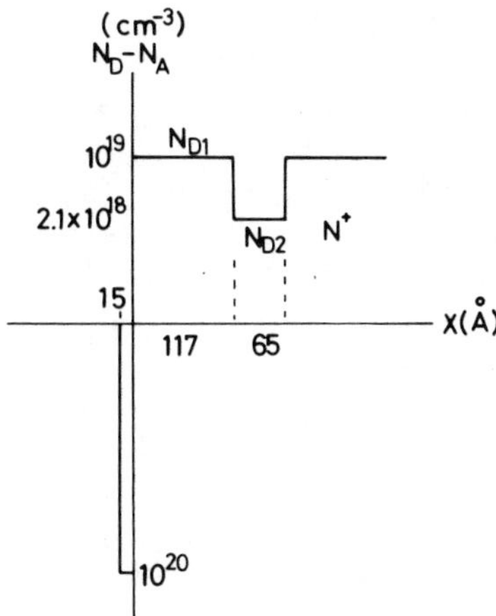

FIG. 31 Calculated doping profile of the p^+-n^+-n-n^+ channel.

determined, N_{D2} is determined by using Eq. (25a). After N_{D1}, N_{D2}, x_1 and x_2 are determined, x_A is determined directly as follows:

$$x_a \geqq \frac{1}{N_A}(N_{D1}\, x_1 + N_{D2}\, x_2) \tag{26}$$

One of the calculated results is shown in Fig. 30. The value of $N_A = 10^{20}\text{cm}^{-3}$, $x_A = 15\text{Å}$, $N_{D1} = 1 \times 10^{19}\text{cm}^{-3}$, $x_1 = 117\text{Å}$, $N_{D2} = 2.1 \times 10^{18}\text{cm}^{-3}$ and $x_2 = 65\text{Å}$ are obtained. The structure shown in Fig. 31 is depleted even at zero bias condition.

The realization of the doping profile as shown in Fig. 30 is difficult by using conventional epitaxy. The photoexcited molecular layer epitaxy introduced in Section III will be useful to realize such a thin and complex doping profile structure. The total active layer thickness is determined by the transit time in a channel and the operation frequency. The treatment of the transit time effect in Section III is also available in case of ISITT. When the transit time effect is appeared, then this device will be called as the ideal static induction tunnel transit time transistor (Nishizawa, 1982).

The noise of the tunnel injection has been confirmed to be very low (Okabe et al., 1968), so the noise properties of ISIT will be low.

V. Conclusion

The detailed and deep consideration of the static induction transistor for the extreme high frequency operation as the millimeter to submillimeter wave region as the ideal SIT was described. The simple predicted performance of tunnel injection as the source injection mechanism named the static induction tunnel transistor will overcome to the ideal SIT by virtue of its inherent high speed by the quantum mechanical effect in the tunnel injection mechanism. The realization of these kinds of transistors and their integrated circuits needs very sophisticated device fabrication technology. It will be realized by using recent GaAs epitaxial technology developed by Nishizawa et al. as PMLE. In the same time, low noise properties of SIT should decrease the lowest power for operation and has already shown operation at 30pW, which is also a very great help for the power delay product which has already shown to be less than 1fJ.

REFERENCES

Abe, A. and Nishizawa, J. (1966). *Int. J. Electronics,* **21** (2), 153–171.

Aiga, M., Higaki, Y., Kato, M. and Yukimoto, Y. (1980). *Dig. Tech. Papers of 1980 Conf. Solid State Devices, Tokyo,* 45–46.

Attla, M. M. (1960). U.S. Patent (Appl. date Aug. 17, 1960) *Semiconductor Circuit Device.*

Bethe, H. A. (1942). *MIT Radiat. Lab. Rep.* **43–12.**

Bozler, C. O. and Alley, C. D. (1980). *IEEE Trans. Electron. Devices,* **ED-27,** 1128–1141.

Cogan, R., Regan, R., Bencuy, I., Butler, S. and Rock, F., (1983). *IEDM Tech. Digest,* 221–224.

Dacey, G. C. (1953). *Phys. Rev.,* **90,** 749.

Esaki, L. (1958). *Phys. Rev.,* **109,** 603.

Kajiwara, Y., Aiga, M., Higaki, Y., Kato, M., Yukimoto, Y. and Shirahata, K. (1979). Proc. 1979 Int. Conf. Solid State Devices, Tokyo; *Japan. J. Appl. Phys. Suppl.* **19-1** 305–308 (1980)

Kato, M. and Kajiwara, Y. (1977). *1977 Spring Meeting of Japan Soc. of Applied Physics,* Paper **24a–13.**

Mead, C. A. (1960). *Proc. IRE,* **48,** 359–361.

Mochida, Y., Nishizawa, J., Ohmi, T. and Gupta, R. K. (1978). *IEEE Trans. Electron, Devices,* **ED-25**, 761–767.

Nishizawa, J. and Watanabe, Y. (1953). *The contract research report to the Nippon Telegraph and Telephone Public Corporation.*

Nishizawa, J. and Watanabe, Y. (1958a). *Science Rep. Res. Inst. of Tohoku Univ. B (Elec. Comm.),* **10** (2), 75–89.

Nishizawa, J. and Watanabe, Y. (1958b). *Sci. Rep. Res. Inst. Tohoku Univ.* **10** (2), 91–108.

Nishizawa, J. (1960). in "Semiconductor devices", pp. 190–194, *Series of the Elementary Course in Electronics Engineering, Tohoku University,* **12,** Kindai Kagaku-sha, (in Japanese).

Nishizawa, (1972). 1972 IEEE International Electron Devices Meeting, Washington, D. C.

Nishizawa, J. and Kitsuregawa, T. (1973). Japanese Patent No. 1038314, *Semiconductor Device,* (Appl. date: May 19, 1973).

Nishizawa, J., Terasaki, J. and Shibata, J. (1975). *IEEE Trans. Electron. Devices,* **ED-22,** 185–197.

Nishizawa, J. and Wilamowski, B. M. (1976). Proc. 1976 Int. Conf. Solid State Devices, Tokyo: *Japan. J. Appl. Phys. Suppl.* **16-1,** 151–154 (1977).

Nishizawa, J. and Wilamowski, B. M. (1977). *ISSCC Tech. Dig.,* 222–223.

Nishizawa, J. and Yamamoto, K. (1978). *IEEE Trans. Electron. Devices,* **ED-25,** 314–322.

Nishizawa, J. Ohmi, T. and Tekeda, N. (1979a). *IEEE MTT Workshop.*

Nishizawa, J. Tamamushi, T. and Ohmi, T. (1979b), IEEE Trans. Electron. Devices, **ED-26,** 1970–1977.

Nishizawa, J. (1979c). Proc. 1979 Int. Conf. Solid State Devices, Tokyo; *Japan. J. Appl. Phys. Suppl.* **19-1,** 3–11 (1980).

Nishizawa, J. (1980a). *Proc. 1980 SRI Conf. Semiconductor Electronics,* pp. 1–22.

Nishizawa, J. (1980b). *Dig. Tech. Paper, Conf. Microwave Solid State Devices, Gdansk, Poland.*

Nishizawa, J. (1982). in *The GaAs TUNNET Diodes,* vol. 5 of *Infrared and Millimeter Waves,* edited by K. J. Button, 215–266, Academic Press.

Nishizawa, J. Tamamushi, T. and Suzuki, S. (1983). in *Semiconductor Technologies 1983,* JARECT Vol. **8** (OHM & North-Holland).

Nishizawa, J. and Kokubun, Y. (1984a). *Proc. 1984 Int. Conf. Solid State Devices and Materials, Kobe,* pp. 1–4.

Nishizawa, J. Abe, H. and Kurabayashi, T. (1984b). *Tech. Group Meeting of IECE of Japan,* **84** (127), 73. (in Japanese), also to be published in the *J. of Electrochem. Soc.,* May (1985).

Nishizawa, J. (1985). In *Atomic Layer Epitaxy on III-V Compounds,* UK Workshop on Atomic Layer Epitaxy, sponsored by the Science and Engineering Research Council (SERC), 25th Feb. 1985 at STL, UK.

Nakamura, T., Yamamoto, M., Ishikawa, H. and Shinoda, M. (1978). *IEEE J. of Solid-State Circuits,* **SC-13,** 572–576.

Okabe, T. Takamiya, S., Okamoto, K. and Nishizawa, J. (1968). IEDM, ibid. *RIEC Technical Rep.* **TR-31,** Tohoku Univ. (1968).

Okamoto, K. and Nishizawa, J. (1966). *Solid-State Electronics.* **9,** 97–103.

Persky, G. (1972). *Solid-State Electronics,* **15,** 1345–1351.

Read, W. T. (1958). *B.S.T.J.,* **37,** 401.

Richman, P. (1969). *IEEE Trans. Electron. Devices,* **ED-16,** 759–766.

Shino, T., Kano, H. Aoki, K. and Ono, S. (1979). *Japan. J. Appl. Phys. Suppl.* **19-1,** 283–287.

Shockley, W. (1952). U.S. Patent 279-0037 (Appl. date Mar. 14, 1952) *Semiconductor Signal Conversion Device.*

Shockley, W. (1952). *Proc. IRE,* **40,** 1289–1313.

Shockley, W. and Prim, R. C. (1953). *Phys. Rev.* **90,** 753.

Shockley, W. (1954). *B.S.T.J.,* **33,** 799.

Shumka, A. (1969). *J. Appl. Phys.* **40,** 438–439.

Shur. M. and Eastman, L. F. (1980). *Electronics Letters,* **16,** 522–523, and also *IEEE Trans. Electron. Devices,* **ED-26,** 1677–1683 (1980).

Suntola, T. (1981). *Proc. of the 5th Int. Conf. Vapor Growth and Epitaxy (ICVGE-5),* 125a–125b.

Suntola, T. (1984). *Proc. Int. Conf. Solid State and Materials, Kobe,* 647–650.

Suzuki, S. (1983). In *CMOS SIT, Report of the ERATO of Research Development Corporation of Japan* (in Japanese).

Teszner, S. and Gicquel, R. (1964). *Proc. IEEE,* **52,** 1502–1513.

Watanabe, Y. and Nishizawa, J. (1950). Japanese Patent No. 205068 (Appl. Date; Dec. 20, 1950) *Semiconductor Devices Including High Resistive Layer.*

Watanabe, Y., Nishizawa, J. and Yoshida, Z. (1951). *Busseiron Kenkyu,* **41,** 96.

Zener, C. (1934). *Proc. Roy. Soc. London,* **A-145,** 523.

Zuleeg, R. (1967). *Solid-State Electronics,* **10,** 449.

INDEX

D

E